Cyber Security Meets National Secu
International Perspectives on Digital Er

Neil Kent
and
Irina du Quenoy

Editors

Cyber Security Meets National Security: International Perspectives on Digital Era Threats

Neil Kent
and
Irina du Quenoy
Editors

St. James Studies in World Affairs

Academica Press
Washington~London

Library of Congress Cataloging-in-Publication Data

Names: Kent, Neil (editor) | du Quenoy, Irina (editor)
Title: Cyber security meets national security : international perspectives on digital era threats | Neil Kent and Irina du Quenoy
Description: Washington : Academica Press, 2022. | Includes references.
Identifiers: LCCN 2022947439 | ISBN 9781680537789 (hardcover) | 9781680537796 (e-book)

Contents

Digitalizing the World: The Era of Invisible Power

Thomas Flichy de La Neuville
(Rennes School of Business)

Introduction

It is clear that the digitalization the world, which interconnects people and equipment, represents a major civilizational transition. It is changing the ways in which power is exercised, partly concealing it from public view. Hackers operate anonymously, attacking websites that collect unwitting users' information.[1] The cybernetic era is therefore making power less perceptible and this, in turn, makes research into this current technological shift crucially important.[2] However, this endeavor is not without risk. There is a danger that we mistake advances in digital technology for the true vitality of a civilization built on the ability to breathe new life into every part of society.[3] Currently, the real and virtual worlds are hybridizing, generating fears around human life being stripped away in favor of technology. This concern is balanced out by hopes that a form of digital humanism will arise. Still, the situation is

---

1 The idea behind all anonymizing systems is to blend various network users' communications together so that they become individually indistinguishable. As a result, it becomes impossible to draw a correspondence between data and identity.

2 Cyberattacks have now become so sophisticated that it is impossible to determine where the perpetrators are located. This makes it difficult to implement aggressive tactics as recommended by the latest cyber strategy white paper.

3 Arnold Toynbee wrote that "if we were to look at the history of war techniques in isolation in Greek history, we would observe continuous progress from beginning to end, before, during and even after the growth period. We would also observe that each step on the road to progress is stimulated by events that are otherwise disastrous for Hellenic civilisation." Toynbee, *L'histoire* (Paris: Gallimard, 1951), 219. Translation from the French text.

difficult, because invisible power is not easy to spot.[4]

Indeed, one of the dilemmas faced by research into cybersecurity is that most data is not accessible to the wider public. This complex field of study can only be deciphered using the technical capabilities of the army or intelligence services. Still, it is possible to overcome this barrier by exploring the sources of information at either end of the chain. Upstream, a small number of philosophers are reflecting critically on the digital world. One example is Bernard Stiegler, who considers that the internet is a disruptive technology, in that digital automation leads to tax avoidance and unemployment.[5] Eric Sadin, meanwhile, defines artificial intelligence (AI) as a kind of rationality that interprets various situations in real time in order to continually propose services and products, a technoliberalism aiming to mould behavior. Lastly, Kave Salamatian sees the internet as a many-tentacled beast with a hyperconnected heart, whose underwater infrastructure provides an indication of the state of digital geopolitics.[6] Downstream, various blogs and websites testify to the current vitality of technology. It is useful to cross-reference technology website Wired.com, which is written for a nonspecialist audience, with French websites InternetActu.net and Reflets.info, which offer a more critical reading of the subject.

Various online tools also enable us to assess the evolution of digital technology, from maps of the underwater cable network that carries 99 percent of internet data to maps of the users of TensorFlow or Shodan.io, which provides an overview of connected devices.[7] It is also possible to approach the gamer-hacker community, which is not opposed to talking about its underground activities (although IT specialists can be reluctant to come into contact with a world quite different from their

---

4 The wider public is entirely unaware of how Google's search engine works.

5 For her part, magistrate and cyber specialist Myriam Quéméner examines how French law is changing in reaction to disruptive digital technology.

6 The Suez Canal and Strait of Malacca are strategic points for underwater cables. China, meanwhile, is only connected to the world via four points.

7 The Marea network is made up of cables no bigger than a hosepipe and was completed in 2017. It links Virginia to Bilbao and currently transmits 75 percent of the world's current online traffic.

own).[8] However, neither upstream nor downstream information sources are location-specific. As a result, monitoring software such as Tadaweb.com and relational mappers such as Gephi.org give us an idea of digital geopolitics in specific places. Counterintuitively, the enhanced imperceptibility of digital power does not tend to smooth out the idiosyncrasies in connected individuals' data. Digitalization thrives by collecting personal data on a massive scale and this, in turn, means breaking societies down into preidentified microgroups. Enclosing groups of individuals into online silos is a prerequisite for effective personalized marketing, of which electoral marketing is an offshoot.[9] Paradoxically, when human societies are digitalized, their identities become more distinct, to the extent that microgroups of connected but single-minded individuals find themselves in opposition to one another.[10] Pay close attention to it and the permanent connectivity generated by all-governing algorithms appears more belligerent than unifying.[11] To understand this, we have to examine the digital revolution's workings so that we can sketch out the potential geopolitical consequences.

---

8 The website Pastebin, for instance, lists the most recent hacking operations. See also Tim Jordan and Paul Taylor, "A Sociology of Hackers," *Sociological Review* 46, no. 4 (November 1998): 757–80. Note that people with Asperger's syndrome (a form of autism characterised by communication difficulties) are particularly likely to thrive in IT professions. Steve Silberman, "The Geek Syndrome," *Science*, January 12, 2001.

9 One of Quantcube Technology's flagship products is Global Macro Smart Data, a real-time predictive platform licensed annually to users. Since May 2013, Quantcube Technology has predicted twenty-one election results with 92 percent accuracy, several weeks before the polls even opened. The start-up notably predicted that the UK would vote to leave the EU in 2016 and that Donald Trump would win the 2016 US presidential election (a fortnight ahead of time in the latter case), as well as the results of France's first round of presidential elections in 2017 and the 2018 American mid-terms.

10 Major digital platforms' most significant means of reducing friction between these different identities is to moderate users' posts. This invisible task was analyzed by Tarleton Gillespie in *The Custodians of the Internet* (New Haven: Yale University Press, 2018), 296. It consumes a huge amount of resources. Global platforms use automated detection tools to moderate content. The task of moderation itself involves flagging up offensive or inappropriate content.

11 Unless, of course, these virtual confrontations threaten the cyber security upon which foreign investments depend.

The Commercial Dynamics of the Digital Revolution

Driven by an ambition to replace human unpredictability with artificial intelligence, the digital revolution uses captology to monopolize consumers' attention and has remained largely untouched by cyber-dysfunctionalities.

Artificial Intelligence: A Trojan Horse Designed to Rob Humans of Their Independence

In the years to come, the development of artificial intelligence will be flanked by the development of 5G and quantum computing. 5G is one hundred times faster than 4G and interconnects people with digital devices, providing the conditions for everything from smart cities to automated environments. 5G is designed for the Internet of Things, such as smart cars and drones, for example. Switzerland has taken an early lead in this area, launching 338 5G masts on April 17, 2019. This technology is not without its risks, however, as it must be relayed every 800 meters. The waves are very high in frequency and noncontinuous. The situation is further complicated by the fact that China has a number of 5G-related patents, with obvious geopolitical implications. In May 2019, US president Donald Trump banned American telecoms networks from buying Huawei equipment. Indeed, the United States has expressed fears about wide-scale espionage and has pressured its allies to freeze out the Chinese company. However, should Huawei and other Chinese businesses be prevented from rolling out 5G capabilities in Europe, the cost to European telecoms operators would be €55 billion and eighteen months of lost time.

On the quantum computing side, new quantum computers (and IBM Q in particular) are rendering current cryptographic protocols obsolete. While the ten or so quantum computers currently in existence are still in the experimental stage, in the next few years, the development of AI will primarily allow it to not only guide consumers' choices but also to reform state bureaucracy.[12] A new science – captology – has been developed to this end.

Captology: The Art of Invisible Manipulation

AI is first and foremost about emotions.[13] Algorithms turn our mental space into code in an effort to capture our attention. The origins of captology reside in the work of Stanford University's B. J. Fogg, who published *Persuasive Technology: Using Computers to Change What We Think and Do* in 2003. Persuasive technologies have been designed by choice architects to nudge people's choices in a certain direction. Rooted in behavioral economics and neuro-marketing, they lend algorithms a governing power. This new economy considers our attention spans to be a rare and precious resource that can be used to a business's advantage when it is focused on information that, in turn, provokes a particular action. Recognizing the potential negative effect of these techniques, Bernard Stiegler has criticized psychotechnology that short-circuits our attention spans, as had Noam Chomsky and Edward S. Herman before him when they denounced the media's role in manufacturing consent. Captology can also builds on artificially generated anxieties, such as the "fear of missing out." This particular fear is largely fueled by certain aspects of modern technology, such as mobile phones and social networking sites, such as Facebook or Twitter, which enable users to constantly compare their profiles. As the world's internet usage grows, a proportion of individuals will become psychologically dependent on being online and suffer from anxiety whenever they are not logged in.

Captology's potential for generating negative outcomes within the context of global digitalization is clear–and yet in the imagination of the global citizenry and security services it is overshadowed by fear of cybercrime, which, however, tends to be overemphasized in the

12 State start-ups could find themselves at the cutting edge of government reforms.

13 For Pierre Bellanger, "software and algorithms are human thought in code. Only a third of our brains is dedicated to rational cognition. If the majority of our thoughts emerge from our unconscious, as soon as networks develop they become unmoored from reason, control and any understanding of our own thoughts. The internet is coding and connecting up our dreams." Translation from the French text.

surrounding discourse.

Cybercrime's Minimal Impact on Global Digitalization

Although cybercrime is becoming increasingly prevalent, especially in developing countries, groups of hackers rarely manage to paralyze organizations or states for long periods of time.[14] In reality, cyberattacks (which are often supported by states) are a sophisticated version of three ancient practices, namely *sabotage*, *espionage*, and *subversion*.[15] Although the costs involved in these operations have plummeted, cybersabotage is still limited in scope. Still, it has been shown capable of doing damage in various ways: it is now possible to blow up a pipeline, break a dam, scramble a radar, delay a nuclear program, shut down a bank, take control of a car or assassinate someone by booby-trapping their mobile phone.[16] Cyber-espionage is the real growth area: today, spies are overwhelmed by the amount of stolen data on offer, rendering them incapable of interpreting all this information shorn from its cultural context. In the world of espionage, attacks are becoming increasingly sophisticated. Certain viruses erase themselves as they go, for instance, making them impossible to trace.[17] They can also mutate, just like biological viruses.[18] As for subversion, while it is easy

---

14 Cybercrime is mostly fueled by developing economies, which have recently become a significant force as far as fraud is concerned. Attacks originating in Brazil continue to rise, making it one of the world's five most prolific cybercrime hotspots. Currently, the most effective hackers operate from Israel and India.

15 The Tor network, which is spreading through the darknet, uses the onion routing technology that was developed by the US Navy in the 1990s. This military technology was mostly financed by the US government to promote democracy around the world. Until 2017, the US government funneled nearly all its funding via the US Navy, the State Department, and the Broadcasting Board of Governors that emerged from the CIA. As a result, major intelligence agencies can reverse data anonymizing processes to uncover Tor users' true identities.

16 Hackers can control the automatic windows and the indicators on a dashboard, or even cause the engine to malfunction.

17 According to Kaspersky Lab's predictions for targeted threats in 2019. The next viral infections will be imperceptible.

18 Viruses use an unstable enzyme to multiply. Because there is no corrective system to call upon, this error remains in the genome. It occurs in about one out of every 10,000 cases. Much variation therefore exists within a viral population.

to start a protest movement online, it is very difficult to maintain in the long run. As a result, sabotage makes it temporarily trickier for the world to go digital, whereas espionage harvests the results of our constant connectivity. Ultimately, only noncommercial subversion is capable of endangering the current digital transition, but it is too infrequently executed to pose any serious kind of threat.

Having examined the digital revolution's workings, we can now sketch out the potential geopolitical consequences. Indeed, as can be seen from the preceding discussion, global digitalization is rooted in market dynamics and enables power to be exerted invisibly upon connected individuals. From a broader standpoint, it is aiding financial capitalism maneuver toward a new international geopolitical configuration, to which we now turn.

Digital Empires vs. Digital Vassal States

Geopolitics in the digitalized world are characterized by three major trends: an erosion of American power, an increase in Chinese power, and competition between the two to digitally colonize the rest of the world.

Eroding America's Digital Power

American power entered the virtual world via an oligopoly: GAFAM. Made up of five major businesses (Google, Apple, Facebook, Amazon, and Microsoft) and founded by former hackers,[19] GAFAM physically stores information.[20] As such, this big data is accessible to the United States and its British relay station. 80 percent of data goes through the United States, and listening stations are positioned close to

19 Examples include Steve Jobs (Apple), Mark Zuckerberg (Facebook), Bill Gates (Microsoft) and Linus Torvalds (Linux).

20 The internet's infrastructure is mainly made up of high-capacity underwater fibre optic cables linked up to terrestrial cables and routers. For Europe, the most important cables are those that link the continent to the UK and, from there, to the US via the Atlantic Ocean.

where underwater cables reach land.[21] The geopolitical implications for control over cables range from fostering dependency among overseas territories far from major population centers,[22] rising Chinese-American tensions,[23] and in other, less-publicized arenas such as Portuguese-Brazilian wrangling for control over the former Portuguese colonies' economic market.[24] Most notably, in response to US dominance other actors have increasingly taken countermeasures: For example, to evade American surveillance, Brazil has built its own cable link to Spain. Similarly, China has built the Sea-me-we 5 cable to connect with the Middle East and, ultimately, Europe. And the Huawei Marine group is laying more and more underwater cables, for instance between Brazil and Cameroon.

Because GAFAM behaves like a state, Denmark dispatched an ambassador to it in 2017. In reality, it is more like a kleptocracy, which is living off stolen data, and whose hybridized belief system borrows from both left-wing libertarianism and right-wing technological determinism.[25] The resulting techno-libertarianism is espoused by visionary, charismatic founders who generate innovative commercial offerings that are always technologically avant-garde. With support from investment funds, GAFAM spent $58 billion on research and

---

21 Historically, underwater cabling helped the British Empire to increase its financial power. The first functional cable was laid in 1851 between the coasts of France and England with the primary aim of telegraphing stock market information. Today, the Seaborn Networks consortium has now started building an underwater fibre optic connection between Fortaleza in Brazil and Wall Street. Known as Seabras-1, the project will eventually link up to African financial markets via South Africa.

22 New Caledonia has been linked to Australia via the Gondwana-1 cable since 2009, guaranteeing it the fastest possible communications. Orange is also planning to link French Guiana to the Caribbean.

23 The United States has responded vigorously to increasingly powerful Chinese investments. In 2013, the American administration thwarted plans to lay a new transatlantic cable between New York and London, to which Chinese firm Huawei was meant to contribute.

24 The SACS cable has linked Angola to Brazil since February 2019, while ELLALINK connects Portugal to Brazil via Cape Verde.

25 GAFAM's system bears historical comparison to the extremely mobile Venetian gondolier company, which was able to pass information shared by travelers to the doge.

development in 2016. They partly owe their monopoly to the brilliant Chinese and Indian minds that staff Silicon Valley. They are still in competition with China, however, which is trying to prevent them from developing any further.[26] To maintain its dominant position in this environment, the United States must speed up its industrial integration with Europe. For its part, if it hopes to reestablish its digital sovereignty, Europe needs to redouble its own efforts and investments in digital technology.[27] If it does not, it will have to accept strategic alliances which reduce it to nothing more than a digital vassal state.

China's Rise to Cybernetic Power as an Indicator of Rare Resources

China's demographic collapse has forced the emerging power to concentrate on technology. The PRC is now hugely connected, and over half of its population has been online since 2017. Four hundred million Chinese people play online games, for example. Chinese internet users are online for about three hours a day, and over half of that time is spent on a mobile device. They browse and watch online videos, abandoning TV screens in favor of more nomadic options. Within this context, as part of its attempt to encourage digitalization while safeguarding against domestic political change, China is seeking to disentangle itself from ICANN, a legal authority based in California that regulates the internet, by boycotting multilateral meetings. Its objective is to draw a strict dividing line between the "Chinese internet" and the "global internet." In an attempt to "cleanse" the internet, the Chinese Communist Party has long sought to strengthen its grip on what it describes as information pollution and "electronic opium."[28]

---

26 There are similarities between China's efforts and the way the United States attempted to block the USSR's forays into the nuclear and aerospace industries in the 1950s.
27 It should be noted, for example, that despite its legal stability and general appeal, France has not been able to create any ground-breaking innovations within the technology realm for the past twenty years due to its rigid labor laws.
28 Internet addicts are viewed as both "deviants" who need their behavior corrected and "sick patients" who require medical help.

From another angle, the Chinese have outclassed the Americans when it comes to AI. AI thrives when data is collected in massive quantities, and China has huge amounts of homogenized data that enables it to outstrip the United States in this realm. The country's "social credit" system, which awards every citizen a certain number of points based on behavior, has enabled it to extract vast amounts of data since it came into force in 2020.[29] Two modes of plundering data are thus going head-to-head: GAFAM pilfers our data on the one hand, while the Chinese government pilfers its own citizens' data on the other. The Chinese have the data but the Americans have the algorithms, which is why it is so crucial to the Chinese that they manage to plunder the latter or attract the very best engineers who will help them catch up. Competition here is fierce and has left behind India, which is only responsible for developing existing programs. For example, in order to maintain its digital independence, China restricts how its rare metals are exported and used, as these will go on to be used in the mobile phone manufacturing process. For Guillaume Pitron, the battle for rare metals is the hidden side of the digital revolution, and in the future, tensions will become particularly high in territories targeted by cyber-colonists.

Major Powers Competing to Digitally Colonize Africa

If the digital colonisation process were modeled on an algorithm, it would include the following steps. First, the colonizer would facilitate a cyberattack on the target state's communications and present itself as the latter's savior. Next, it would connect the country to the internet, secure its strategic networks and capture local cyber elites by rolling out a master's degree program in digitalization. It would then pillage the country's data, take over the consumer market and, finally, engage in online electoral marketing to maintain its grip on its conquered market.

This is precisely what is happening in Africa, where connectivity is growing rapidly. In 2018 alone, 35.2 percent of Africans were using the internet, compared with 16 percent in 2012. In this context, China has

29 This system is based on the questionnaires used by certain American insurance companies.

taken a very original stance toward the continent. First of all, Chinese businesses have provided the continent with technological tools at very competitive prices. For instance, Shenzhen-based electronics manufacturer Huawei has been operating in Africa since 1999. In 2013, it partnered up with Microsoft to conquer the African smartphone market and is now outperforming the Californian giant in this sector in Africa. Huawei's ambition is now to sell premium telephones to an African middle class looking for enhanced services, further positioning China as a leader in this market. There are, however, significant differences between countries, from North Africa to the Sahel region, as the Chinese are not the only cyber colonizers in town.[30]

For example, the British are also taking an interest in this promising market and organized a cyber-forum in 2018. France, too, has been seeking a piece of the pie. In 2019 Orange set up a specialist cyber defence subsidiary in Morocco: Orange Cyberdefense Morocco officially opened for business in Casablanca on April 16, 2019. A master's degree in cyber defense has been set up in partnership with the Université Polytechnique Hauts-de-France in Valenciennes to attract African talent and launch their careers. France also has a long-established presence in Libya as, in 2007, Amesys—a French subsidiary of Bull—sold its Eagle program to the Gaddafi regime to track its opponents. After changing its name to Advanced Middle East Systems, the company then sold a similar system to Egypt known as Cerebro. In the Sahel region, practices are changing, as internet cafés offer faster internet speeds rather than access to computers, suggesting rapidly progressing digitalization. The French army ran several cybersecurity courses, including in Nouakchott, Mauritania, from September 10 to 13, 2018, and in Niamey, Nigeria, from January 29 to February 3, 2018.

The Middle East: Digital Islands and Rich Pickings

Finally, it is worth considering here another region where digitalization is taking on increasing geopolitical importance–the Middle

---

30 For example, Algeria – considered the least secure country—had to call in specialist cyber police officers to monitor its baccalaureate exams in 2017.

East. The latter has two digital "islands," Iran and Israel, which represent opposing yet mirroring forces. On the one hand, Israel provides 7 percent of the world's cybersecurity, and its industry benefits from high levels of investment that aim to protect a territory stripped of its strategic importance, while also improving its ability to export innovations.[31] On May 6, 2019, Israel sent out a warning to hackers operating outside its borders. After one cyberattack in particular, the Israel Defense Forces bombed a building in the Gaza Strip that was sheltering Hamas hackers. This has not prevented Israel itself from recruiting hackers keen to offer it their services for a hefty fee.[32] The field is growing, as exemplified by three former Israeli intelligence officers who founded XM Cyber in 2016 to operate in this particular field. For its part, and in opposition to Israel, stands the digital island of Iran, a country with equal levels of creativity but lacking in Western investment.[33] Among other things, Iran appears to have been behind cyberattacks on certain British banks in December 2018. The viruses used—Shamoon 1, 2, and 3—also targeted petroleum infrastructures in the Gulf's oil monarchies.[34]

Yet, Iran and Israel are not the only actors in the Middle East whose digitalizing activities have an impact outside their own borders. In Turkey, the nationalist Cyber-Warrior Akıncılar group has hacked anyone or any organization deemed opposed to the interests of Turkey and Islam.[35] Saudi Arabia, meanwhile, has enjoyed technical support from certain Israeli cybersecurity firms, which have made use of their position to gather data on the behavior of the country's elites or opponents to the regime.[36] Still, despite the protection offered by these

---

31 $800 million in 2017.
32 Hackers are highly coveted. In fact, Dubai has even organized a hacking fair!
33 The word "algorithm" was invented by Persian mathematician al-Khwarizmi in the ninth century.
34 Cybercrime can sometimes be a proxy for military action. This was the case for these Iranian cyberattacks, which were designed to relieve pressure on the country during the embargo.
35 They hacked into *Charlie Hebdo*'s website, for instance, in 2011.
36 In 2017, representatives of the Herzliya-based NSO Group Technologies held a series of meetings in Vienna and at least one Gulf nation, during which a $55 million contract was signed to supply Saudi Arabia with its famous Pegasus spyware. On the

Israeli companies, Saudi Arabia is now the most frequently targeted Middle Eastern country when it comes to cyberattacks, which mostly aim to steal information from private and public institutions' information systems.[37]

Conclusion

The growth in digital power offers a very imperfect reflection of the real clout of the states under discussion here: GAFAM's rise obscures the United States' geopolitical decline, while China's technological power conceals its demographic fragility. Conversely, in both Africa and the Middle East, hacking operations designed to steal digital or financial resources reveal the weaknesses of supposedly powerful forces.

Social digitalization rates are therefore a better reflection of the resources that are immediately available rather than their future potential. The internet has become a bitterly contested space among competing economic powers. The aim of this competition is to effectively privatize a space that is temporarily free for users to explore. Permanent connectivity, which devours instant data and interconnects objects and reified human beings, bypasses anyone who refuses to accede and labels them as suspicious. In the future, we can predict that states and individuals will generate a fog of incorrect data in an act of defiance designed to shield themselves from view.[38] Finally, digitalization is

---

opposition side, Jamal Khashoggi and Omar Abdulaziz wanted to set up a cyber-opposition movement.

37 The lack of local training and a general lack of awareness of the dangers of cyberattacks have left the kingdom exposed.

38 Voice cloning is now a reality. Using a single minute of recorded audio, Lyrebird and Wave Net can digitally recreate an individual's voice to have him or her "say" whatever they want. Similarly, Stanford University has demonstrated facial control in which a person's facial expressions are recorded live but edited by an actor, with a computer instantly reproducing the latter's movements yet with any target person's face. Such deepfakes can attack individuals, organizations, and states. University College London has developed its *My Text in Your Handwriting* program to accurately reproduce a person's handwriting from a single sample. A talking robot named Luka mimics the characters from the TV series *Silicon Valley*, recycling dialogue from the first two seasons. The robot responds to questions by generating

throwing up unexpected geopolitical changes. Ultimately, it will sweep away repetitive jobs and temporarily concentrate power in innovative territories, while making it possible for states and businesses to identify opinion leaders using relational mapping. Lastly, it will enable military chiefs to order robots to kill targets using AI-powered facial recognition. We are entering into the era of hybrid civilizations.

---

new phrases based on the models provided to it. In 2022, it is believed that populations in developed nations will read more fake news than genuine information. Automatic text, image and audio generators could contaminate the entire human communication network. 8.5 percent of Twitter accounts are already run by bots.

The "Strategic Corporal": Facing the Cyber Threat in French Armies and Homeland Security Forces

Antoine-Louis de Prémonville (officer in the French Army and independent researcher in History and Geopolitics)

Introduction

The young men and women joining the ranks of the French Army and homeland security forces every year have a lot in common with their fellow citizens. Their habits, leisure, and tastes are those promoted in the everyday life. Contrary to the cliché, French servicemen don't live on the fringes of society. Most of them surf the Web and use social networks to keep in touch with their family and relatives while serving in France or overseas. In fact, nothing in particular seems to protect them against exterior threats; on the contrary, their activities make them valuable targets. Indeed, the French Armies and other security services have tended to adopt the newest trends and technologies for communication and self-promotion. Nowadays, the internet is available in every boot camp and most officers daily send e-mails daily for internal communication.

Granted by the French constitution, servicemen have the same rights as any other citizens —even if they obtained the right to vote after women, on August 17, 1945! However, among the very specificities of the military, they have a duty to exercise restraint regarding political, philosophical, and religious opinions, in order not to harm the e-reputation of the Armies.

As the very last wardens of the Republic, the armed forces should not allow servicemen to express personal points of view in the name of the institution. The dangers of this are clear. For example, even though some service members served loyally as members of Parliament under the Third Republic, the government at the time feared the influence of conservative officers, most of them Catholic, Bonapartist, or royalist. That is the reason why an 1875 law disqualified them from standing as candidates for parliament election. Admittedly, Field Marshal de Mac Mahon openly endorsed plans for royalist restoration and General Boulanger had planned a military coup. Loyal but openly reluctant in his support for the republic, Field Marshal Lyautey was far too critical as a secretary of war. And, of course, we cannot forget Field Marshal Pétain and the Vichy regime.

Yet, today, expressing discord with the official line seems inconceivable. Recall that the army's nickname is the "Big Mute" (*la Grande Muette*)! Even if servicemen can be elected to Parliament, they must leave office for the duration of their term.[1] Along with the preservation of the reputation of the armed forces, the discretion of servicemen on social networks is thus a professional imperative.

In order not to compromise both the confidentiality of operations and their own security in a context of frequent terrorist attacks against representative of the state, servicemen must obviously keep some secrecy regarding their own engagement. It is not only an elementary professional skill; it is a primary need, for the consequences are often serious. Here, I will talk examine how servicemen's lack of discretion on social networks could harm the entire security organization. (1) First, I will discuss the e-reputational damage. (2) Then, I will focus on the risk of compromising missions, which is a threat that sometimes leads to (3) tragic and deadly endings.

Reputational Harm

In a famous article published in 1999, U.S.M.C. general Charles C. Krulak insisted on the role of "strategic corporals" in modern wars, that is to say, in asymmetric conflicts mixing peacekeeping and

---

1 "Laetitia Saint-Paul, seule militaire élue députée, fait son entrée à l'Assemblée," *Europe 1*, June 24, 2017, https://www.europe1.fr/politique/laetitia-saint-paul-seule-militaire-elue-deputee-fait-son-entree-a-lassemblee-nationale-3370739.

humanitarian aid in an over-mediatised atmosphere. His aim was to demonstrate that a low-rank private could seriously compromise the mission—and the campaign as a whole—if a wrong decision with negative consequences were to be mediatised.[2] Subsequently, French retired colonel and academic Michel Goya tried to propose a mathematical formula for Krulak's concept. Namely, he wrote, "We could formulate an equation of the strategic corporal with the following elements:

- F representing the initial folly
- R representing the portrayal of the folly
- S representing its spreading in the media
- C representing the unfavourable context.

Note that C resonates with F. Indeed, an unfavourable context leads to folly, either deliberately because of provocative actions or simply due to an atmosphere of distrust."[3]

In April 2004, human rights violations committed by US Army and CIA personnel against detainees in the Abu Ghraib prison in Iraq came to public attention. The scandal tarnished the reputation of the US Army, whose mission was supposed to promote democracy and human rights. Condemned both within the United States and abroad, the affair is a textbook case of the strategic corporal's negative effect in cases of misconduct. Indeed, the infamous photos have remained visible on the Web, years after the prosecution and condemnation of black sheep servicemen.

Devastating for the reputation of the armed forces, dubious video with dead or submissive persons rapidly go viral because of the easy sharing on websites like YouTube. Similarly harmful, explicit references to renegade political regimes made by individuals wearing a uniform generally force top officials to condemn them. Haunted by its painful

---

2 Charles C. Krulak, "The Strategic Corporal: Leadership in the Three Block War," *Marines Magazine*, January 1999.

3 Michel Goya, "Le caporal stratégique ou peut-on confiner la connerie?" *La Voie de l'épée*, June 1, 2020, https://lavoiedelepee.blogspot.com/2020/06/le-caporal-strategique-ou-peut-on.html.

history and extremely vigilant when it comes to ethical issues, the Bundeswher was shaken by a "desecration of skulls" scandal in 2006.[4] In the same vein, German, Australian, or French armies—among others—have been confronted with leaked photos and videos showing servicemen praising the Third Reich.[5] However, praising horrors is not the only way for servicemen to tarnish the reputation of the institution they serve in. Numerous apparently "benign" cases underline how powerful and hurtful speeches and opinions by personnel may undermine the armed forces' image. For example, both the *gilets jaunes* ("yellow jackets") and the anti-gay marriage movements have been praised by an unknown number of members of the French Ministry of Defense. As individuals, they benefit from the liberty to demonstrate, provided they are dressed in civilian clothing. However, any incident or violence they are involved in during a demonstration is generally mentioned in the media, as is their membership in a public institution. Obviously, this does not mean that there are plenty of servicemen who belong to militant groups attacking the police and causing extensive damage. At the same time, during arrests of rioters and looters, some suspicious passers-by have been known to work for Defense. Even though they could be found innocent—black sheep do exist, let's not deny it— the career consequences are hardly ever neutral.[6]

There are other types of reputational risks besides servicemen holding objectionable opinions or taking part in violent riots. Between

---

4 Thomas Rid, "Les photos du caporal stratégique. Comment les nouveaux médias changent la guerre," IFRI, November 2006, https://www.ifri.org/en/publications/publications-ifri/articles-ifri/photos-caporal-strategique-medias-changent-guerre.

5 "New Photos Link Elite German Soldiers to Nazi Emblem," *Deutsche Welle*, November 2, 2006, https://www.dw.com/en/new-photos-link-elite-german-soldiers-to-nazi-emblem/a-2223277; Dan Oakes, "Australian Soldiers Flew Nazi Swastika Flag from Vehicle in Afghanistan; PM Says Diggers' Actions 'Absolutely Wrong,'" *ABC*, June 13, 2018, https://www.abc.net.au/news/2018-06-14/photo-shows-nazi-flag-flown-over-australian-army-vehicle/9859618; "Trois militaires photographiés faisant le salut nazi," *Le Figaro*, April 2, 2008, https://www.lefigaro.fr/actualites/2008/04/02/01001-20080402ARTFIG00283-trois-militaires-photographies-faisant-le-salut-nazi.php.

6 Jean-Dominique Merchet, "Une lettre du colonel Ruffier d'Epenoux," *L'Opinion*, June 18, 2013, https://www.lopinion.fr/blog/secret-defense/lettre-colonel-ruffier-d-epenoux-1164.

2011 and 2016, the scandal over Louvois, the defective software dedicated to pay the wages of Army personnel, was directly responsible for massive unpaid wages, provoking delicate personal financial situations. In 2012 only, members of the army experienced an estimated 465 million Euros in unpaid wages. While the Ministry of Defense did recognize the situation as a failure and acknowledged its shameful consequences, committees of servicemen spouses and partners popped up on social networks to denounce the unbearable circumstances into which they were thrust. The breaking of a kind of omertà suddenly enabled the media to address the consequences for families and the amount of public funds wasted in this dramatic failure.[7] Due to the outrage people expressed over the scandal, the Army had no choice but to jettison Louvois and replace it with a new software. For our purposes, even though the burden of responsibility rested with the politicians who had decided to implement the defective software, the scandal damaged the Army's reputation. Considering the slow political response and the hidden pressures some servicemen experienced from their hierarchy to silence some spousal committees, many soldiers, NCOs, officers, and their relatives did interpret the situation as a form of disrespect.

Compromising Missions

In addition to the risk of tarnishing the reputation of the Armed Forces, misuses of social network sites are very likely to compromise mission safety. What are we talking about, precisely? In any operation, one of the key success factors is keeping the highest possible level of confidentiality. When the enemy knows the very moment, the place, or the composition of the detachment, he is then capable to respond appropriately, adapting his own modes of action to foil ours. Indeed, knowing the enemy is essential to anticipating his modes of action. That is the reason why all Armed Forces have an intelligence service dedicated to the understanding of the enemy. However, the Armed Forces are not the only actors that try to adopt the opposite point of view.

---

7 Jacques Monin, "Louvois, le logiciel qui a mis l'armée à Terre," *France Inter*, January 27, 2018.

The military capabilities of the enemies the French army has been fighting since the 2000s are inferior by far to French standards. However, it would be a serious mistake to consider these terrorist armed groups as being unable to seek out information on our capabilities and modes of action. Certainly, their intelligence cells have limited means. Until we are shown proof to the contrary, the Islamic State or other Islamic armed groups in the Sahel lack high-tech intel capabilities such as satellites, electronic warfare, air support, or Special Forces operators. On the other hand, it would be risky to assume they are unimaginative. They have discernible capabilities of diverting certain technologies from their primary purpose or adapting obsolete military equipment to meet new needs. Those so-called barefoot thugs dressed in rags and driving rusty pickup trucks know how to design unpiloted armed aerial vehicles, send encrypted messages, and infiltrate government departments and agencies.

Considering their ability to inflict casualties and, what is more, to keep on fighting for years despite the limited means at their disposal and the tremendous losses experienced, we need to face that they have true professional skills. Moreover, maybe should we admit, too, that this enemy knows us better than we know him. Many terrorists were born and raised in Europe. Western pop culture influences anyone watching the television or listening to the radio anywhere in the world. That is the other side of soft power's double-edged sword. Hence, it would very be risky to underestimate the existing threat to the point where no safety precautions are taken. The ability to surf the Web and connect to social network sites is easily accessible worldwide. Moreover, accessing personal data nowadays is all the easier given that people share them with sometimes unknown third parties on the Web. Some people publicly share so many photos, videos, and posts on Facebook, Twitter, Instagram, YouTube, and so on, that one can easily establish a file on them. Unfortunately, some members of the armed forces (or any other public agency) misuse the online networks. Were an agent to forget that the Internet is not a forum for discussion about their missions but a means to stay in touch with relatives, he would compromise his mission. Such a situation is not hypothetical. One of the most famous cases happened in 2010, within the Israeli army. Specifically, an Israeli private published a post on Facebook revealing details of an upcoming mission in the West Bank: "On Wednesday we are cleaning out the village of Katana (nr Ramallah) – today [an] arrest operation, tomorrow an arrest operation and then please god, home by Thursday." Other servicemen reported the post to the unit's commanding officer, who decided to cancel the mission for fear that operational security had been breached. In 2008 another Israeli soldier was jailed for uploading sensitive information onto Facebook.[8] Considering the number of social networks users in the French armed forces, such a possibility could arise there, too. A 2014 survey pointed out that three out of four army service members, most of them noncommissioned, were active social networks users. Only 11 percent had given up their online profiles [upon joining the army], whereas 15 percent (one out of three being officers) had never had any.[9] Hence, in 2012 the French Ministry of Defense released a *Guide to the Correct Use of Social Networks*. Meanwhile, cyber security experts regularly give conferences in boot camps and military academies to warn the personnel about the dangers posed by social networks to operations abroad and in the homeland.

More recently, new threats from social networks have arisen in the cyber battlefield. Deception is a well-known mode of operation aimed at influencing the enemy's manoeuvres. Now, it is being conducted on the Web. For example, some opponents of the French army in the Sahel spread obvious fake news to foster a popular anti-French movement. According to these fake news items, so-called friendly fires incidents by Barkhane operators would frequently involve its African allies. Others pretend that the French support armed terrorist groups. Clearly unfair and partisan, these lies are massively shared, "liked," and commented upon.[10]

---

8 Simon McGregor-Wood, "Facebook Details Force Israeli Military to Cancel Operation," *ABC News*, March 4, 2010, https://abcnews.go.com/International/facebook-details-force-israeli-military-cancel-operation/story?id=10006343.

9 Michel Sage, " Les militaires dans l'espace public numérique," in *Guerre, armées et communication*, ed. Eric Letouturier (Paris: CNRS éditions, 2019), 124.

10 Laurent Lagneau, "Sahel: Sur les réseaux sociaux, la force Barkhane est visée par une campagne de fausses informations," *Opex360*, December 5, 2019,

Rumours about so-called organized fanatics serving undercover in the Armed Forces serve to spread suspicion within the units. Although some former members of the French armed forces have in fact joined the Islamic State to conduct holy war (jihad),[11] most Muslim soldiers are loyal to the French state. Indeed, a famous fake piece of news, which has been shared for years on several websites, deals with a so-called mutiny onboard the aircraft carrier *Foch* in 1999. According to this rumour, Muslim sailors, fed up with the bombing campaign against Kosovo (a now Muslim country) they were participating in, attempted to revolt. However, this never actually occurred.[12]

For all that, some potential Islamic terrorists and violent radicalized individuals have certainly been detected for years in the French armed forces and police. We must admit the enemy is competent in using new media tools and establishing communication strategies, capable of expressing himself ever more undisguised in the most efficient ways possible on the Web. Indeed, the design and production of some propaganda films and magazines do conform to Western standards.[13] Webzines like *Dabiq*, *Rumiyah*, *Inspire*, and *Dar al-Islam* were issued irregularly in 2014–2016 in several languages, including French. They provided starter packets to potential terrorists who could obtain technical and ideological material just by reading them. The defeat of the Islamic state as a territorial entity has not put an end to their sharing information or propaganda on the Web or the relevant digital support media.

---

http://www.opex360.com/2019/12/05/sahel-sur-les-reseaux-sociaux-la-force-barkhane-est-visee-par-une-campagne-de-fausses-informations/.

11 The Centre d'Analyse du Terrorisme has identified approximately thirty former French military who have either joined the theater of operations on the side of terrorist organizations, primarily in the Iraqi-Syrian zone, or have participated in terrorist operations in France. See Manon Chemel, "Les militaires français et le djihad," Centre d'Analyse du Terrorisme, December 2019, http://cat-int.org/wp-content/uploads/2019/12/CAT-Militaires-Djihad.pdf.

12 Guillaume Daudin, "De fausses preuves d'une infiltration islamiste de la Marine nationale," *AFP Factuel*, October 18, 2019, https://factuel.afp.com/de-fausses-preuves-dune-infiltration-islamiste-de-la-marine-nationale

13 Alexis Marant, *Studio de la terreur*, Canal+, 2016.

A Deadly Threat to French Security Forces

The terrorist threat to the homeland has made military men anonymous in their own country. Nowadays, wearing the uniform has been banned outside the service. Names are erased and faces are blurred to prevent identification during media coverage. As a result, soldiers have vanished from everyday life and from peoples' mind. With the exception of the seven to ten thousand soldiers of operation Sentinelle who patrol the streets to ensure the security of sensitive areas—mostly religious sites—French people can no longer see soldiers dressed in battledress or uniform in a bakery or shopping in a supermarket. Some retired military officials and anonymous senior officers do denounce what they consider a retreat in front of the enemy. Such an argument is not without merit. However, we must admit that over the past decade, members of the armed forces and other security services have suffered heavily from terrorism. Let us just remind three famous cases in France illustrating how service men have been targeted in today's war against terrorism on home soil. In all three cases, cyber security played a key role.

The Merah Case, 2012

Mohammed Merah was a chronic offender well-known to the French homeland security services. A radicalized Muslim, he had traveled several times to Pakistan and Afghanistan to pass through traineeships in terrorist boot camps. Although he pretended—and was eventually believed—to be traveling to the region as a tourist, some homeland security analysts continued to monitor his activities. On March 11, 2012, he set out finally to criminally act out in pursuit of jihad, targeting servicemen. He was able to identify his first victim easily: the latter had posted an advertisement with a far too explicit description on a widely known e-commerce site. The serviceman, an airborne NCO named Imad Ibn Ziaten, wanted to sell his motorbike; in the ad, he explained that he had no choice but to get rid of it as he was frequently away from home. In conclusion, he wrote, "I am a serviceman." Merah feigned interest in the bike and arranged an appointment with Ibn Ziaten; upon meeting him, he denounced Ibn Ziaten as a traitor to Islam and killed him. Three days later, he attacked a group of servicemen who were passing by in a street at the gates of their military compound in Montauban. Two soldiers were killed and another one was seriously injured. On March 19, 2012, he finally attacked a Jewish school, killing one adult and three children.

Identifying the perpetrator thanks to the Internet Protocol (IP) address he had used while connecting to the e-commerce site, the police launched a vast manhunt. Merah, who had dug himself in at home and resisted a thirty-two-hour siege, was finally killed in an assault. Five members of an elite police force were injured during the action. Afterward, several jihadist armed groups claimed responsibility for Merah's attacks. Even murkier, the unclear nature of the relationship between Merah and the intelligence services has generated controversies about so-called security gaps and unproven theories of collaboration.

The Slaughter of Magnanville

The second case happened on June 13, 2016, when a twenty-five-year-old terrorist named Larossi Abballa slaughtered a police officer couple in their home in Magnanville. The couple's three-year-old son was spared thanks to the intervention of another radical Muslim, with whom Abballa was chatting on Facebook live. As he was filming the killings to claim his membership in the Islamic State, Aballa questioned himself about what to do with the kid. The terrorist, who had been convicted three years earlier of assisting a terrorist plot, was shot a few hours after the murders by an elite police squad.[14] The reason why he decided to kill those two persons remains unknown. However, it seems he did not choose the couple randomly. Analysis of his cell phone revealed he had surveilled the area several times. Moreover, in October 2017, while searching the home of a woman suspected of complicity in a terrorism case, the police seized a USB pen on which they found the names of 2,626 police officers. Also included was a list of the names,

---

14 "Ce que l'on sait de Larossi Abballa, le meurtrier d'un couple de policiers dans les Yvelines," *France Info*, June 14, 2016, https://www.francetvinfo.fr/faits-divers/terrorisme/ce-que-l-on-sait-de-larossi-abballa-le-meurtrier-d-un-couple-de-policiers-a-magnanville_1499095.html.

identification numbers, and units of all the senior members of a police union (dated to 2008). All these police officers served in the homeland security services. As to how this information had leaked—the sensitive data had been shared with all the agents mentioned in the list and with the union itself, making the breach of information all the easier.[15] Indeed, the suspect had been housed by a close friend's mother who was an NCO police officer and union representative who had converted to Islam. The police interrogated her regarding the Magnanville slaughter, as she did know people about to participate in jihad.[16] However, due to a lack of evidence, no charge could be laid against her.

Harpon Case, 2019

On October 3, 2019, Mickaël Harpon, a civil servant working at the Intelligence Bureau of the Paris Prefecture, assassinated three policemen and an administrative agent in the courtyard of this institution, whose purpose includes preventing terrorism. There was no doubt in classifying the killing as a terrorist attack, since its perpetrator's aim was to "« kill unbelievers."[17] For the very first time, France faced a "blue on blue" terrorist attack, a friendly fire attack carried out on purpose by a member of the force.[18]

---

15 "EXCLUSIF. L'identité de 2 626 policiers de la DCRI aux mains d'une radicalisée," *Le Point*, April 10, 2018, https://lepoint.fr/societe/exclusif-lcs-noms-de-2-626-agents-du-renseignement-aux-mains-d-une-radicalisee-10-04-2018-2018-2209549_23.php.
16 "Une clé USB avec les noms de plus de 2000 fonctionnaires de police retrouvée chez une jeune femme radicalisée," *France Info*, April 4, 2018, https://www.francetvinfo.fr/faits-divers/terrorisme/policiers-tues-a-magnanville/une-cle-usb-avec-les-noms-de-plus-de-2000-fonctionnaires-de-police-retrouvee-chez-une-jeune-femme-radicalisee_2699414.html.
17 Simon Piel, "Attaque à la Préfecture de police: Une recherche Internet faite par Mickaël Harpon accrédite l'hypothèse terroriste," *Le Monde*, February 26, 2020, https://www.lemonde.fr/societe/article/2020/02/26/attaque-a-la-prefecture-de-police-une-recherche-internet-faite-par-mickael-harpon-accredite-l-hypothese-terroriste_6030930_3224.html.
18 In 2012, the French army experienced a "green on blue" attack caused by an allied Afghan soldier, who killed unarmed soldiers practicing sport at Guan FOB, Afghanistan.

In the Harpon case, the most serious thing is that he had access to classified data, as he was a technician responsible for the maintenance of computers. In spite of his evident radicalization, the terrorist could consult lots of personal data concerning Prefecture employees. Confidential data as well as jihadist propaganda were found on his USB pen. While subsequent investigations could not confirm there were leaks of confidential data, this was due only to an absence of direct evidence.[19] In the aftermath of the killings, intelligence experts have publicly expressed concerns about how a well-known radical Muslim, who attended extremist mosques, could had been granted such a level of security clearance.[20] A special commission of inquiry has proposed thirty-five measures to prevent such a risk in the future.[21]

Conclusion

A misunderstanding of the cyber threat causes the human failures at the root of most security breaches in the armed forces and other security services (injury of reputation, compromising, attacks on members' lives). The recent cases underscore that one need not be a senior hacker or employ high-tech to cause losses. Increasing controls may produce no results if no one observes the procedures. However, a powerful and long-term awareness raising campaign aimed at informing servicemen on the cyber threat is starting to yield encouraging results, as a change in mentality can be observed. Indeed, cyber security has become quite inevitable in the everyday life of servicemen.

---

19 "Attaque à la préfecture de police: Pas de trace de fuite de données, la clé USB toujours analysée," *Europe 1*, October 9, 2019, https://www.europe1.fr/societe/attaque - a - la - prefecture - de -police-pas-de-fuites-de-donnees-la-cle-usb-toujours-analysee-3924453.

20 Jean Chichizola, "Affaire Harpon: La commission dénonce 'une faille majeure,'" *Le Figaro*, May 3, 2020, https://www.lefigaro.fr/actualite-france/affaire-harpon-la-commission-denonce-une-faille-majeure-20200305.

21 "Attaque de la PP: La commission d'enquête fait 35 propositions pour répondre aux 'failles,'" *L'Express*, October 6, 2020, https://www.lexpress.fr/actualite/societe/attaque-de-la-pp-la-commission-d-enquete-fait-35-propositions-pour-repondre-aux-failles_2127862.html.

To face these new threats, France has taken various measures. First, its intelligence services have had to adapt to the e-revolution. Within the intel world, new departments and cells dedicated to very specific know-how have been set up. Along with the external intelligence service (DGSE) and the Military Intelligence Service (DRM), the Department for Intelligence and Defense Security (DRSD) seeks specifically to protect all MoD servicemen and cells. Dedicated to counterintelligence, its aim is to identify opponents' intentions and to neutralize any internal and external threats which may lead to hostile acts from organizations, groups, or individuals. DRSD agents investigate and protect. They identify weaknesses and contribute to bolstering security. Among others, one of the key missions of the DRSD is to mentor troops on counterintelligence (securing data, preserving secrecy, avoiding reputational damage) prior to any tour of duty overseas. Every year, several servicemen are punished by the military high command, and sometimes by the courts, for judicial misconduct because of offenses that intel services managed to identify or prevent.

Nowadays, most electronic communications are encrypted. The armed forces use dedicated email and computer networks that are supposed to remain unbroken and hermetic to hackers' assaults. Today, any platoon leader frequently uses encrypted software to communicate within his own unit. Moreover, the MoD has been hiring hundreds of computer science specialists, software designers, data analysts, hackers, and other kinds of "geeks" to join ancient and brand-new units such as the 44th and the 54th Signal battalions, the 805th Signal company (created in 2016), which stands alongside the 785th electronic warfare company in Brittany, as spearheads for cyber security and counterintelligence. Even if much still must be done to compete with superpowers on the virtual battlefield, France has remained credible in how it understands the strategic issues. Hopefully, considering its indigenous know-hows and the renowned multinational defence groups (Dassault System, Sagem, Safran, Thalès, etc.) based in France, the issue is rather more "we will do it" than "we can do it."

Cybersecurity and Cybercrime in Nigeria: The Implications on National Security and Digital Economy

Babayo Sule (Department of Political Science, Federal University of Kashere Gombe, Gombe State Nigeria); Bakri Mat (School of International Studies, Universiti Utara Malaysia); Usman Sambo (Department of Public Administration, Yobe State University, Damaturu, Yobe State Nigeria); Mohammed Kwarah Tal (Department of Political Science, Federal University of Kashere Gombe, Gombe State Nigeria); and Muhammad Aminu Yahaya (Department of Public Administration, Gombe State University Gombe State Nigeria).

Abstract

This study examines the implications of cybercrime and weak cybersecurity defense for Nigeria's national security and digital economy from the perspective of nontraditional discourse within international security debates. The research adopted a qualitative approach to data collection and analysis, with both primary and secondary source data employed. The primary data involved in-depth interviews with selected informants from the relevant agencies in Nigeria and the available governmental policy documents on cybersecurity and cybercrime, including conversations with security personnel, academics, senior officials from security research institutes, members of private institutions and government agencies in communication sectors. The secondary data consist of books, journals, and internet sources. Our analysis reveals that cybercrime is flourishing in Nigeria undetected, affecting its critical national infrastructure, causing prolonged terrorism affecting national security and the safety of national environment due to weak cybersecurity capability. The article recommends that Nigeria adopt measures toward strengthening its cyberspace and digital environment to prevent crimes and promote national security and the digital economy.

Introduction

The shift in the new paradigm of security discourse that emerged during the era of Cold War and beyond stresses the need to pay attention to nontraditional or nonmilitary threats to domestic and international security (Buzan & Hansen 2009). Indeed, the emergence of widespread security challenges such as marine pirates, flood, drought, diseases, terrorism, environmental hazards, and cybercrimes necessitates the rethinking of the foundation of contemporary global security (Buzan 1983, 4). Cybersecurity in particular poses an unrelenting challenge, threatening the existence of mankind and infrastructures across nation-states. Countries are making serious efforts in protecting their cyberspace from cyberattacks and cybercrimes in the face of threats and vulnerabilities that could cause billions of dollars' worth of loss in properties, cash theft, and the collapse of critical national infrastructures (Fischer 2009). Cybercrime is on the rise globally (Buchanan 2016), affecting individuals, companies, organizations, and governments with damages worth billions of dollars (Antonucci 2017).

Within this context, Nigeria is a developing country battling threats to national security on numerous fronts, both traditional military threats and nontraditional challenges such as hunger, diseases, erosion, sea piracy, drought, flood, terrorism, and the explosion of cybercrime both within and outside its borders. Recognition of this latter threat led to the adoption of a policy document on cybersecurity by the Nigerian government in 2015, a comprehensive policy statement detailing the provisions and efforts of the government aimed at achieving a safer digital environment. Many scholars (Schjolberg & Ghernaouti-Helie 2009; Reveron & Mahoney-Norris 2011; Schjolberg & Ghernaouti-Helie 2011; Singer & Friedman 2014; Mazurczyk, Drobniak, & Moore 2016; Maurer 2018) have pointed out that internet and cyberspace are global and pose multiple vulnerabilities to all nation states, and in fact, studies on Nigeria (Frank & Odunayo 2013; Olayemi 2014; Omodunbi, Odiase, Olaniyan, & Esan 2016; Ajiji 2017; Chukwuma & Mogom 2018; and Mohammed, Mohammed, & Solanke 2019), though scanty on the subject matter, suggest that Nigeria is as vulnerable as other countries in the world to these threats.

The Nigerian case, however, is particularly important in terms of the study of cybersecurity and cybercrime due to its unique set of circumstances. The country is the most populous state in Africa, and by extension, the biggest Black nation on earth. It is the tenth-largest oil producing state and the seventh most populous country in the world. The explosion of internet access in Nigeria should be of prime concern for both policymakers and key players in the world of cybersecurity, as vulnerability in the Nigerian context is always imminent. Yet, comprehensive analysis of Nigerian cybersecurity issues is woefully inadequate, a gap this article seeks to address. In addition to providing an overview of the situation we offer policy suggestions for the Nigerian government on how to improve on the existing cybersecurity infrastructure in the country.

Literature Review

In this section, we discuss some of the empirical and theoretical issues related to the subject matter of study. These include (1) the emergence of cybersecurity as a central global concern; (2) the evolution and character of the global digital economy; (3) the emergence and patterns of cybercrime globally and in Nigeria; (4) policies and regulations related to cybersecurity; (5) Nigerian national security; and (6) Nigerian policy response to cybersecurity concerns and cybercrime.

Cybersecurity in the Global Context

In the contemporary era, the rapid expansion of internet communication technologies and digital activities, particularly the proliferation of cellular phones, has created a new security concern, as large quantities of stored data have become the target of hackers and criminals (Rayes & Salam 2019). Further, financial transactions carried out in the internet arena serve as a clear avenue of vulnerability. With over five billion people in the world possessing mobile phones and over four billion having access to the internet, cybersecurity has become an inevitable and pressing concern (Global Web Index 2019). As cyberattacks by nation-states among themselves and by individual criminals against both states and public users have proliferated, a national security paradigm shift has emerged in intellectual discourses (Reveron et al. 2011). In the specific case of Nigeria, twenty-first-century statistics on mobile owners, internet users, and the digital economy suggest that cybersecurity must be taken as a serious concern by both government and private individuals.

Digital Economy

The digital economy and indeed, digital activities such as e-commerce, e-governance, and e-learning, have been undergoing transformation for several decades. Countries have been moving toward digital space for ease of life since the end of World War II, a phenomenon that dates to the invention of the first computer (Severo 2019, 165). With the emergence of internet services, digital activities began to proliferate in the 1980s (Noam, 2019). The contemporary explosion in internet systems across the globe and the creation of communication gadgets such as mobile phones, laptops, iPad, tablets, and other gadgets that use internet services has effectively succeeded in digitalizing the world (Schwanholz & Graham, 2018).

In terms of commerce, by the twenty-first century, many businesses had adopted technology aimed at innovation and improvement of customer services and operations, often with positive results in terms of profitability and ease of operations. The advantages of conducting business online are such that in 2018 more than 42 percent of business globally were conducted in this manner (retrieved from https://www.slideshare.net/sap/99-facts-on-the-future-of-business-in-the-digital-economy-2018). At the same time, enterprises that have done so face the risk of unauthorized use and malicious attacks, including theft, the destruction of intellectual property, abuse by insiders, and illegal use of information, which is tantamount to the loss of data reliability and confidentiality and affects the levels of trust between enterprises and their users/customers (Shalhoub & Al Qasimi, 2010).

In an environment where the digital economy is characterized by the physical separation of the buyer from the seller, the separation of the buyer from the product, and the risk of cybercrime, enterprises must find ways of initiating and developing cyber relationships with customers in sustainable ways, of which the best is establishing a trusted digital space. (Shalhoub & Al Qasimi, 2010). The importance this point is underscored when we consider that most countries in the world are currently going digital in virtually all aspects of national importance, such as e-governance, e-commerce, e-voting, e-transaction, e-learning, and other vital aspects of human endeavor in the society. Digital trust and security in this case take on paramount importance and demand the attention of policymakers.

Emergence and Patterns of Cybercrime

Cybercrime emerged historically in the 1970s and 1980s during the Cold War era. During that period, most patterns of cybercrimes revolved around nation-states engaged in the ideological battle and involved attacking the computers and data of rival countries. Important data related to nuclear proliferation, intelligence movements, and foreign policy were the subject of hacking by professionals employed by the relevant governments. At the same time, individual criminals were also actively hacking data, targeting critical national infrastructures and financial organizations for the purposes of stealing of money and sabotage (Kshetri 2019; Mordi 2019). However, in terms of its pervasiveness, cybercrime became far more extensive, dynamic, and pervasive after the collapse of the USSR.

Cybercrime is believed to have generated an amount of money worth \$1.5 trillion in 2018 and \$2 trillion in 2019 across the globe through major hacks and thefts from websites and organizations. The annual total amount (that we know of) realized from cybercrimes consists of \$860 billion from illegal online markets, \$500 billion from trade secrets and IP thefts, \$160 billion from data trading, \$1.6 billion from crimeware; and \$1 billion from ransomware. In the United States, cyber frauds led to losses of $25 billion in 2010 alone; in Britain and Australia, losses in 2015 amounted to 3.6 billion British pounds and AU$94 million. These crimes have largely involved multinational corporations and governmental information, with major examples of data breaches in 2018 including Yahoo ($3 billion), Under Armour ($150 million), eBay ($145 million), Equifax ($110 million), Facebook ($87 million), JP Morgan Chase ($76 million), Paystation ($77 million), Uber ($57.6 million), and Home Depot ($56 million) (Re-Hashed 2019).

Turning to Nigeria now, similar to other developing countries, the level of development in internet and cyberspace in the 1970s and 1980s in that country was minimal. Even though it was an appendage of capitalist states during the Cold War, cybercrimes in Nigeria during this time were practically nonexistent. Beginning with the 1990s, however, Nigeria joined the retinue of states facing nonmilitary security threats, with cybercrime front and central. For instance, if Africa as a whole lost $3.5 billion in 2017 to cybercrimes, the Nigerian portion of this sum was $645 million, by far the largest (Kshetri 2019). A year later, it was reported that Nigeria lost $800 million (N288 billion) collectively to cyberattacks in 2018 (Azeez 2019). More broadly, a 2019 report disclosed that Nigeria has lost on average N127 billion ($328,842,878 million) annually to cybercrimes in recent years (Ohwovoriole 2019). Furthermore, Nigeria has the reputation of being the global headquarters of cybercrimes, due to frequent daily incidences linked to the country (even though Statista data in 2017 did not even mention Nigeria among the top twenty countries in terms of the highest incidence of cybercrime) (Mordi 2019). Finally, in addition to these glaring threats posed by cybercriminals to the Nigerian economy, the Nigerian Communication Commission (NCC) reported in 2018 that there were about 9 million unregistered lines in the country that were used by mobile owners; some of these were clearly being used for criminal activities such as terrorism, rural banditry, communal violence, kidnapping and robbery, posing a serious threat to national security.

Policies and Regulations on Cybersecurity

The need for policies on cybersecurity is anchored in the recognition that there has been a shift in global governance toward e-governance and the security implications arising therefrom. The 2009 cyberattacks on South Korea and United States present a good case for concern in this regard (Andreason 2012, 77). According to Pelton and Singh (2015), cybersecurity laws should target four major areas. The first is personal and family protection, which consists of protecting stocks, bonds, bank accounts and credit cards, Medicare accounts, desktop computers and Wi-Fi, personal cell phones, and family records. The second aspect involves cybersecurity policies for smart phones and free Wi-Fi that operates through mobile phone security and access. The third aspect is protecting vital infrastructure through the application of satellites with cybersecurity implications. Finally, the fourth is strict government regulation and monitoring of cyber activities. The reality, however, is that the development of such policies has been haphazard at best.

In an attempt to minimize the rate and severity of cybercrimes, various international organizations have undertaken the responsibility of pushing toward cyber law development in both the developed and developing countries. A leader in this respect, the International Telecommunications Union (ITU) launched the Global Security Agenda in 2007 and formed a High-Level Expert Group to investigate the issues and develop proposals for long term strategies (Shalhoub & Al Qasimi 2010, 18). The ITU in particular promulgates laws under its auspices that have some applicability to information warfare attacks that use the electromagnetic spectrum or international telecommunications network (Greenberg, Goodman, & Hoo 1998, 7).

On the state level, governments have adopted antihacking laws to help promote cybersecurity and attempted to approach the issue within the framework of the internationally recognized right of nation-states' rights to defend themselves when attacked, in accordance with article 51 of the UN Charter (Guiora 2017, 63). The Convention on Cybercrime, initiated by some European countries, is the only international agreement in existence at the global level in the area of cybercrime. It is also open for membership to countries outside of Europe, being signed for example by the United States, Canada, Japan, and Australia. The convention comprises three key areas: substantive law, procedural requirements, and international cooperation. However, some critics argue that these laws are outdated and have failed to help effectively protect either private or government computers (Kosseff 2017). Further, as observed by Maurushat (2013), the challenges of laws related to cybercrime are many, including but not limited to the issues of freedom of expression/freedom of speech, copyright, tort of negligence, defamation, illegal telecommunications interception (surveillance), privacy law, data protection, and data breach notification.

National Security

The traditional thinking on national security was predicated on the assumption that state actors are the major players in national and international politics (Buzan 1983, 2). A new conceptual paradigm depicts that national security and indeed, international security, is more complex and dynamic in the post–Cold War era because of the emergence of nonstate actors as powerful players in national and transnational politics (Buzan & Hansen, 2009, 9). Indeed, the failure of the balance of power that had reigned after World War II highlighted the importance of collective security in the international system (Johari 1997, 215). To this end, a number of solutions have been proffered, of which perhaps the most prominent is the deterrence theory as advanced by Realist theorists such as Henry Kissinger (Taylor 2012).

Generally speaking, the national security policy of any state must identify threats domestically and contain them accordingly before expansion toward the global level. Every state may be involved in permanent warfare; control over subversion must be through domination and natural leadership, which is attainable through armed forces (Chomsky & Herman 1979). At the same time, in the present-day national security must address more than just military threats, to include climate conditions, social settings, political structure, the economic basis of the country, among other aspects, with cybersecurity prominent among them. With this context in mind, we now turn briefly to a discussion of national security issues in contemporary Nigeria.

Nigerian National Security

Nigeria's national security policy emphasizes domestic protection, African unity, peace, and the promotion of global peace and security through cooperation, noninterference, and respect for the dignity and value of mankind. Since political independence in 1960, Nigeria has adopted many measures to ensure that the country would protect itself adequately while extending its influence at the African level and globally (Yakubu 2012, 12). Of these goals, the fundamental objective is the protection of the lives and properties of all the citizens and other inhabitants of the country. This is meant to be achieved through building social cohesion, economic prosperity, and mutual coexistence between different groups. Ostensibly, the Nigerian defense strategy is anchored in the understanding the centrality of the recognition of the dignity of mankind, personal well-being, and prosperity before other issues (Alkali 2010, 10). However, the failure of policymakers to in fact pay adequate attention these values and ideals created a huge security gap, which manifested in 1967 through the civil war and later in recent years through terrorism, communal clashes, political violence, rural banditry, kidnapping, robbery, and cybercrimes.

Currently, Nigeria faces many security challenges, both internally and externally. Within the country, the volatile political environment and fragile social settings emanating from multireligious, multiethnic, and regional politics have contributed to security dilemmas, exacerbated by a large contiguous geography, porous borders, and economic problems. The Anglophone and Francophone dichotomy is another security concern, in which Nigeria is surrounded by four major West African countries with a French colonial past, creating a wide gap in terms of communication, external allegiance, and cooperation (Yakubu 2012, 14). More recently, Nigeria has come to face a new security threat, namely, terrorism, with the emergence of the Boko Haram, which shields itself under the umbrella of religion to commit heinous crimes against mankind in the northern part of the country, particularly the northeastern enclave. Within this context, Sule et al. (2019) have suggested that the major focus of the Nigerian security apparatus should pivot toward a nontraditional approach based in cybersecurity and intelligence gathering to address nontraditional/nonmilitary threats.

Nigerian Policy Response to Cybersecurity and Cybercrime

In Nigeria, the Cybercrimes Prohibition Act of 2015 is the major policy instrument meant to boost cybersecurity and prevent cybercrimes in the country. The policy document reiterates unequivocally that the major aim of the act is the prevention of damage to critical national infrastructure (part II, sec. 3 and 4). The act describes those offenses that are considered to be cybercrimes and punishable by law (part III, sec. 5 to 36), including:

- offenses against critical national information infrastructure;
- unlawful access to a computer;
- illegal registration of a cybercafé;
- system interference;
- interception of electronic messages, email, or electronic money transfers;
- tampering with critical infrastructures;
- willful misdirection of electronic messages;
- unlawful interceptions;
- computer related forgery or fraud;
- theft of electronic devices;
- unauthorized modification of computer systems;
- network data and system interference;
- fraudulent electronic signature;
- cyber terrorism;
- exceptions to financial institutions posting and authorized options;
- fraudulent issuance of e-instructions;
- fraudulent reporting of cyber threats;
- identity theft and impersonation;
- child pornography and related offenses;
- cyberstalking or cybersquatting;

- racist and xenophobic offenses;
- conspiracy;
- importation and fabrication of e-tools;
- breach of confidence by service providers;
- manipulation of ATM/POS terminals;
- spreading a computer virus;
- electronic card–related fraud and use of a fraudulent device or attached emails and websites (Cybercrime Act, 2015).

The punishments stipulated for the above identified cybercrimes offenses vary but are related in terms of their severity. Some are punishable with ten-year imprisonment without an option of a fine while some risk up to fifteen years without an option for a fine. Others are punishable with a three-year imprisonment or an option of a fine amounting to seven million Niara.

The Case Study and Methodology

As of 2018, the Global Web Index reported Nigeria's population to be 203.6 million, with 169.2 million (83 percent) having mobile phone connections. 89.49 million (42 percent) are internet users and 27 million (13 percent) are active social media users. The digital growth indicators in 2020 from that of 2019 indicate an increase of 5.1 million (2.6 percent) million in Nigerian population. These figures indicate an important threshold in the study of internet and digital communication in Nigeria.

Theoretical Framework: National Security of Developing Countries Model by Alagappa (1987)

This work adopts and expands the on the "National Security of Developing States" model postulated by Muthiah Alagappa (1987), presented and discussed below. We did so for three reasons. The first has to do with the chain of values and clear illustrations that are depicted in the diagram for easy grasp of the concepts involved, namely, national security and possible vulnerabilities and threats. Secondly, previous studies on our case have used other theories, such as political realism, collective security, foreign policy analysis, and other related frameworks, but none of them refer to this framework, clearly demonstrating the innovative nature, contribution, and new ideas proposed by this study. Thirdly, the theory as it stands leaves room for future theory building. Originally developed in 1987, the theory identified some internal and external threats to national security of developing countries at a time when cybersecurity and cybercrime in developing countries were practically nonexistent. But its postulations left the door open to explore threats in the political, social, economic, and environmental areas, making it easy for researchers to lean on the original postulations and build on the theory for further development of knowledge in the field of social sciences.

Figure 1: The National Security of Developing States: A Framework of Analysis

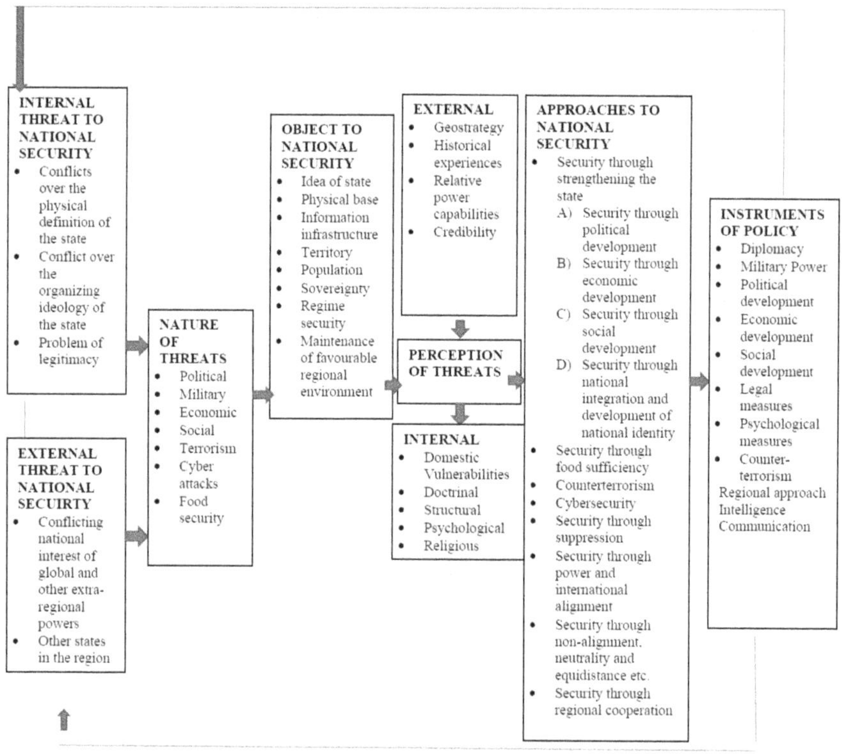

Source: Muthiah Alagappa (1987). Expanded by the researchers.

The above model indicates that there are, esentially, two major threats to national security: internal and external threats, with both being possibly either traditional or nontraditional. In the internal case, there may be social incoherence and other vulnerabilitites that might lead to armed groups—which is a traditional threat—while at the external level there might be territorial and national interest issues that might set one country at conventional military conflict with another. From the nontraditional perspective, issues such as the economy, geography, social settings, or sea and land rights can be potential conflicts that might lead to insecurity internally and externally for developing states like Nigeria. For example, case of Boko Haram here falls within both the traditional and nontraditional threats and internal and external security threats that Nigeria is facing. The nature of threats can be economic, military, political, or, as we propose in this research, related to terrorism, food insecurity, or cyber attacks. Thus, as identified by Alagappa (1987), any serious developing country that seeks absolute national security must pay attention to both internal and external security and also must give consideration utmostly to both the trational and nontraditional military threats.

The framework advanced by Alagappa (1987) is suitable for this study because it has the leverage of providing an all-round explanation of national security of countries like Nigeria, that is, accounting for both traditional and nontraditional threats. Other theoretical frameworks discussed in the above section can explain either the political angle or the economic perspective of cybersecurity and cybercrime. However, the theory we lean upon is all-encompassing in terms of both theoretical and empirical analysis. For instance, looking at the figures of users of mobile phone and internet in Nigeria and examining the loss of billions of dollars in cybercrime, one can easily use this framework to establish and illustrate that the phenomena are related to national security. The trillions of Naira deposited in Treasury Single Account (TSA) by the government and the hundreds of billions used monthly for the payment of salaries under the Integrated Personnel Payroll Information System (IPPIS) present a node of vulnerability if the Nigerian cybersecurity system is hit. In our work, we rely on empirical evidence related to cyber communication by criminals and terrorists in building the theoretical apparatus. In essence, our empirical findings strongly support the theory's suggestion that nontraditional security threats to countries like Nigeria can be as strong as traditional ones.

Our work is inspired, certainly, by that of others. For instance, Cavelty & Wenger (2019) opined that there is a great deal of effort within the international relations discipline aimed at situating its analytical discourse toward an inclusive focus on cybersecurity as a part of global security concerns. Cavelty et al. (2019) stress that cybersecurity is connected to political security because state actors are dealing with state and nonstate actors on the international scene whenever there are threats and vulnerabilities of attacks from a perceived hostile enemy. Cavelty et al. (2019) presented three key drivers that influence cybersecurity politics and research in the field. They include politics (divided into domestic politics and international politics), science (consisting of academic debates and institutionalization), and technology (involving key events and development and use). One of the areas that this research is interested in is domestic and international politics of Nigeria with regard to cybersecurity, and of particular relevance to us is that Cavelty et al. (2019) found it expedient to expand the rationale behind why research of this nature is undertaken from the theoretical perspective.

There are other studies (Koops 2016; Aniekan & Afolabi 2017; Diogenes & Ozkaya 2018; Kremling & Parker 2018; and Kshetri 2019) similar to that of Cavelty et al. (2019) on the nature and necessity of undertaking cybersecurity research and cybercrime protection from the perspective of politics. These studies suggest strongly that any serious country and indeed other organzations as well as individuals today must have a serious provision and concern for securing their personal belongings, data, and environment against political attacks via cyberwarfare. The Nigerian case is especially pertinent because of the shortage of adequate research on the political implications of cybersecurity in the country from the vantage point of national security. Research of this nature should be of paramount importance not only for policymaking but also for setting the foundation of future research on cybersecurity and politics in the Nigerian contexts and by expansion, in Africa. Finally, we should note an interesting work (World Bank, 2017) that emphasizes more the digital aspect of the question: namely, the World Bank posits that the extent to which consumers and organizations are going digital makes it imperitive for countries to secure their cyberspace.

The theoretical and empirical contribution of our research supports the conclusion that the assumptions and strategies presented by the likes of Alagappa (1987), Cavelty et al. (2019), and Kshetri (2019) for identification of cybersecurity as a nontraditional political security threat that requires counteraction should be analyzed and digested by researchers and policymakers in the area of study. To reiterate, we settled ultimately for Alagappa's model because of its permeability, comprehensiveness, and wider horizon in embracing all the elements and variables discussed in the study, such as national security, cybersecurity, cybercrime, digital economy, policy responses, and collective security in general.

Data Collection and Analysis

Primary source data came from interviews with some selected participants drawn from different categories, as well as from the policy document on cybersecurity adopted by the Nigerian government and other reports from relevant agencies. A total number of thirty interviewees were selected, to be sorted in four categories coded as category A, B, C, and D. In category A, eight military or paramilitary security personnel were selected from different agencies directly dealing with security issues. A senior security officer in Nigerian military was selected based on accessibility and possession of quality relevant data on the subject matter of study; another senior officer from the Nigerian Police Force was chosen based on accessibility and possession of valuable information. Other interviewees in this category included a senior officer from the Department of State Security, a senior officer from the Nigerian Immigration Services, a senior officer from the Nigerian Prison Services, a senior officer from the Nigerian Security and Civil Defense Corps, a senior officer from the Nigerian Custom Services, and a senior officer from the Nigerian Federal Road Safety Corps.

In category B, ten senior officials were chosen from institutes and agencies handling security issues, including two senior officials from the National Institute for Policy and Strategic Studies (NIPSS) Kuru Jos, two senior officials in the National Defense Academy (NDA), two senior officials from the Nigeria Police College Wudil Kano, two senior officials from the Ministry of Defense, and two senior officials from the National War College Abuja. For its part, category C comprised academicians, wherein six professors were selected from three universities based on the proximities of the researchers. Two are professors in political science with specialization in security and strategic studies from a reputable national university, two are professors of computer science with specialization in cybersecurity and digital environment, and the last two are professors of sociology with specialization in criminology from a reputable national university.

In category D, six members of private institutions that deal with cyber activities were selected randomly and interviewed. These comprise a senior bank official in Abuja in one of the commercial banks, two senior officers from the telecommunications companies operating in Nigeria, a senior official in the Nigerian Communication Commission (NCC), a senior official from the Nigerian Technology Development Agency (NITDA), and a senior official from the Federal Ministry of Communication. All of them were selected and interviewed in Abuja, where their head offices are located.

There were three criteria used to identify the informants to be interviewed for this study. The first was the identification of stakeholders who are closely linked and related to the subject matter of national security, cybersecurity, and cybercrime and digital economy. The researchers were instructed by various agencies and ministries to consult staff or units on cybersecurity in security parastatals and other ministries. They were directed by senior officials there to consult the most informed in the unit on the subject matter of cybersecurity. The selected informants were interviewed based on the condition of anonymity and confidentiality. The second criteria used was the accessibility. For instance, academicians who were consulted were chosen based on accessibility in addition to their possession of in-depth knowledge on the subject matter; the same was true for some government ministries and agencies. The third criteria used was based on Sharan's postulate (2009) that the minimum number of interviewees for a qualitative research project should be either four or five and that the maximum should be thirty, depending on the availability of the resources and informants with quality information.

While the questions were prepared separately for each category, some general questions were posed to all four categories of interviewees. Semistructured questions were designed to enable for flexibility and free flow of information from the informants. The method used for gathering the data was written questionnaires organized in semistructured sessions, adapted to the different categories of the informants. Some of them were interviewed through phones, some responded through email, and others through WhatsApp.

Finally, as mentioned, the other primary sources of data included the government policy document on cybersecurity enacted in 2015 by the Nigerian government, together with various reports by national and international agencies on cybersecurity in Nigeria. Secondary sources consisted of books, journals, and internet sources relevant to the topic at hand. The data obtained were grouped into themes and analyzed using content analysis in which the previously existing literature was analyzed in relation to the findings of this work and the application of the theoretical framework for conclusion and findings.

Discussion and Findings

This section discusses the major findings from the fieldwork, contextualizing it within the existing literature and the adopted theoretical framework, which enabled the identification of the knowledge gap and contribution to knowledge. The data are discussed in thematic form under three major themes: perspectives on challenges and prospects of cybersecurity in Nigeria, the effects of cybercrimes on Nigeria's national security, and implications of cybercrime for the digital economy.

Perspectives on Challenges and Prospects of Cybersecurity in Nigeria

We have previously noted the rapidly increasing number of internet and computer users in Nigeria, a population that is vulnerable to cybersecurity breaches and cybercrime. The potential for negative outcomes has been further escalated with the emergence of social media usage in the country, to include emails, Facebook, WhatsApp, Telegram, YouTube, Instagram, and many other technologies. In addition to data presented earlier, we should note that digital growth indicators from 2019 to 2020 show that mobile connections increased by 12 million (+7.7 percent), the number of internet users grew by 2.2 million (2.6 percent), and active social media users increased by 3.4 million (14 percent). These data substantiate the conclusion that Nigeria has been going digital rapidly in recent years: for example, from 2013 to 2019 the number of internet users increased from 51 million to 93 million, as seen in figure 2.

Figure 2: Number of Internet Users in Nigeria from 2013 to 2019

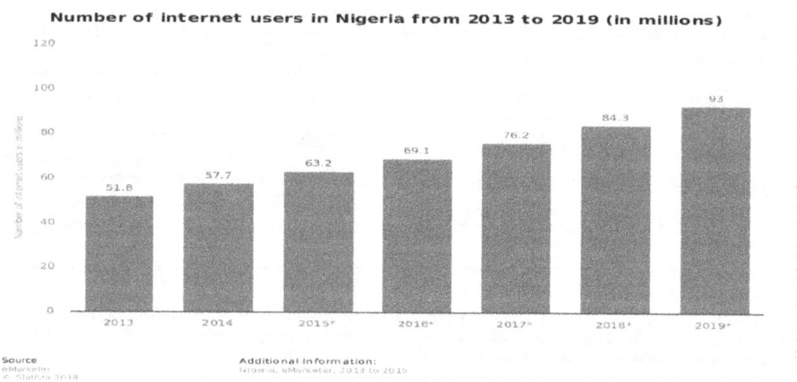

Source: Adopted from Statista 2019.

Another source from the Nigerian Communication Commission (NCC) disclosed the figure of internet users in Nigeria as 111.6 million in December 2018, an increase of more than 10 million over Statista's data. In any event, the increase in the use of social media also contributes to the vulnerability and exposure of the private users and government agencies, organizations, and corporations to cybercrime.

The figures above imply that there is a clear need in Nigeria for allocating a huge sum of money and adopting a strong policy on cybersecurity in order to minimize vulnerability and effectively combat related crimes. Indeed, this was the justification for the adoption of the Cybersecurity Act in 2015. On the other hand, Nigeria faces a serious challenge in the cybersecurity sector because of poor infrastructure, low levels of digital know-how and low-level information transmission. The 2018 Cybersecurity Index concisely pinpointed Nigeria's vulnerabilities in this regard (table 1).

Table 1: Nigeria's Cybersecurity Index 2018

Rank	National Cybersecurity Index	Digital Development Level	Difference
45	44.16	35.86	8.30

Source: Global Cybersecurity Index 2018.

And yet, not all of our sources viewed the prospects of and challenges to cybersecurity in Nigeria in identical ways. We summarize and present their views here in tabular form (table 2).

Table 2: Responses of Informants on the Prospects and Challenges of Cybersecurity in Nigeria

Category	Informants	Prospects	Challenges
A	1	Strong rules and regulations	Lack of enforcement
	2	Improved awareness	Information flow
	3	Existence of cybersecurity laws	Lack of adherence to rules
	4	Global cooperation and interconnectivity	Borderless crimes
	5	Increase users and malware protection	Anonymity aspect
	6	Global awareness forum and policies	Weak digital know-how
	7	Increased protection	Poor management of

		through policies	cybersecurity
	8	Low level of internet users	Vulnerability
B	1	Emergence of sanctions on cybercrime	Borderless nature of internet
	2	Availability of policies on cybersecurity	Vulnerabilities in the cyberspace
	3	Increased information flow	Low level of infrastructure
	4	Emergence of programs on cybersecurity	Limited number of experts
	5	Technical support from advanced countries	Suspicion and lack of digital trust
	6	Bridging the knowledge gap on internet	Low level of information flow
	7	Establishment of cybersecurity laws	Failure of enforcement by authorities
	8	Digital protection through policies	Anonymous and external threats
	9	Training and modern equipment	Low technical support in operation
	10	Points for identifying suspicious activities	Corruption and mismanagement
C	1	Increasing the horizon of security studies	Inadequate professionals and experts
	2	Existence of specific policies and sanctions	Lack of adequate awareness
	3	Global awareness and technical support	Vulnerability and human rights
	4	Increasing conscience of the public	Global nature of cyberspace
	5	Strong policy and performance	Low level of protection
	6	Spying on criminals for detection	Infringement on privacy
D	1	Low level of digital activities	Lack of knowledge on the field

2	The presence of regulations	Fragile security situation
3	Training of professionals for cybersecurity	Lack of political will
4	Management information flow system	Lack of qualified manpower
5	Digitalization policies by the government	Lack of awareness of the threats
6	Resistance to fraudsters and cybercriminals	Increase number of internet users

Source: Field Survey from interviews conducted between July to December 2018 with the selected informants by the researchers.

In addition to the summary above, there are one or two views that are revealing and should be expressed here. In an interview, one of the informants narrated that "Nigeria is tilting towards cyberspace like other world countries but there is a huge challenge that should awaken the policymakers towards particularly securing the cyberspace through an effective cybersecurity. This is because Nigeria lost N288 billion in 2018 alone to cybercrime which is alarming and a threat to the internet users in the country" (Interview with an informant in category A on October 27, 2018, in Abuja at 11:30 am). From a different perspective, another informant revealed that "the weak cybersecurity space in Nigeria disclosed how vulnerable the country. Nowadays, we are no longer battling with terrorism alone but cyberterrorism because even the terrorists themselves were using the cyberspace through the anonymity doctrine to undertake their heinous crimes" (Interview with an informant in category C on September 18, 2018, at 9 am).

Effects of Cybercrime on National Security in Nigeria

Cybercrime has now become a major security headache across the globe because of its nature, which is cross-border, anonymous, and very difficult to detect or prove. In Nigeria in particular, the cybercrime indices indicate a harmful and vulnerable situation, with the country listing among the ten top countries in the cybercrime index as reported by US Federal Bureau of Investigation. The report revealed that the United States ranks first, United Kingdom second, Nigeria third, followed by Canada, Romania, Italy, Spain, South Africa, Russia, and China. Apart from the above, the interviewed informants revealed different views on the effects of cybercrime on national security in Nigeria. The views are summarized and presented below in a tabular form (table 3).

Table 3: Summary of Informants Views on the Effects of Cybercrimes on National Security

Category	Informants	Responses
A	1	It creates room for terrorism
	2	It escalates insurgency
	3	It creates fear and spread of undetected crimes
	4	It leads to the spread of insecurity in all parts of the country
	5	Resources in billions are lost annually
	6	Kidnappings and other crimes are flourishing undeterred
	7	Vital national information is accessed without authorization
	8	People are not safe in the country, including security personnel
B	1	Critical national infrastructure is at risk
	2	Individuals are affected in their private dealings
	3	Terrorists are utilizing cyberspace for coordination of attacks
	4	Internal and external fraudsters are defrauding Nigerians undetected
	5	Security personnel are finding it difficult to handle insurgency
	6	Counterinsurgency is facing setbacks because of the use of the internet
	7	Criminals are extorting billions of naira through e-fraud
	8	The economy is affected through cybercrimes
	9	The social setting is suffering from crimes undertaken using internet

	10	The emergence of cellphones allows for cybercrimes and insecurity
C	1	Cybercriminals use smartphones and internet for their activities
	2	Terrorism is increased by cyberspace through internet communication
	3	External treats exist trough e-attacks on digital businesses
	4	Hacking of organizational data leads to stealing of billions of naira annually
	5	The national economy is dragged backward because of threats
	6	The digital economy is sluggish due to fear of cybercrimes
D	1	Cybercrimes lead to increased terrorism which affects national security
	2	Online businesses are affected by cybercrimes via mutual suspicion
	3	Vital governmental information is accessed by unauthorized persons
	4	National security planning is at risk of hacking by criminals
	5	Use of private cellphone and emails exposes individuals
	6	Individuals are tacked and targeted through cyberspace

Source: Field survey from interviews conducted between July to December 2018 with the selected informants by the researchers.

In an interview, one informant underscored a crucial point, namely that "critical national infrastructures are at great risk in the country such as payment platform like IPPIS and Treasury Single Account as well as many other segments that are using the internet for delivering important services to the nation. The repercussions are that once Nigeria by either advertent or inadvertent accident found itself under cyberattacks, the damage would be unimaginable" (Interview with an informant in category B on November 11, 2018, at 8:30 pm). Furthermore, another informant disclosed that "the global trend for national security is shifting towards cyberspace and collective security is now moving globally towards securing the cyberspace which is more important. Criminal activities that can threaten national security of a country like Nigeria such as terrorism, cyber fraud, looting of national treasury and lack of national databank for emergencies are possible if the cybersecurity of Nigeria is not given adequate attention that it deserves as other world countries are doing today" (Interview with an informant in category D on December 19, 2018, at 11:00 am). Theoretically, the Alagappa model is useful to strengthen the analysis here because the activities of hackers and terrorists as indicated above show that the insecure cyberspace in Nigeria has greatly exposed the country to criminal activities that affect national security.

Additionally, during fieldwork the researchers themselves noted a daily incidence of an average of two mails per day via their emails, where they are targeted for cybercrimes and fraud. Some of the emails bore European, American, or Arabic names, claiming to come from the United Nations and other reputable international organizations; the content of the email would ask the recipients to assist in the transfer of millions of dollars from various world destinations into Nigeria. The targeted individual would be asked to help by providing his account details for the transaction and in return he would allegedly receive half of the proceeds of the transfer. In other mails, the targeted individual would be congratulated and asked to forward his details as the winner of the World Health Organization's annual charity, or a prize given by the United Nations, UNICEF, or other related agencies. All the informants who were consulted as interviewees for this study also attested to the same experience. Sometimes it is a direct call from some unscrupulous elements claiming that you must send your account number and Bank Verification Number (BVN) for rectification of an account issue. Once sent, you will start receiving alerts that you are withdrawing money, your account hacked, and the money stolen. "In essence, once your email is online and you are a frequent user of internet and emails especially academicians who published articles with their details of addresses and other organizations that place the details of their staff on net, you are vulnerable and exposed to these cybercriminals" (Informants in category B, views expressed during an interview on September 17, 2018, at 9:30 am).

In another view, some of the informants also narrated that cybercrimes are carried out through the use of computers and smartphones, leading to national insecurity. For instance, terrorists in Niger-Delta who steal oil and bomb the oil wells and pipelines communicate through phones and emails; similarly, the Boko Haram terrorists carried out their insurgency using internet communication such as computers, YouTube to release video messages, phones, and other means of modern communication. The lack of a proper mechanism put in place by the government to detect the criminals is causing a serious security breach and a potential threat to national security as is now being witnessed where the entire country is enmeshed in one form of insecurity or the other such as Boko Haram in the northeast, armed bandits and kidnapping in the northwest, a Farmers-Herders conflict in the north central region, militants' activities in Niger-Delta, kidnappings in the southeast and armed robbery in the southwest.

Implications of Cybercrime on Digital Economy

Globally, there are around three and a half billion internet users and five billion mobile phone users. Ninety-nine percent of the world's population lives in a place that is covered by a mobile-cellular network and 84 percent of people on the planet live somewhere covered by mobile broadband networks (Graham 2019, 3). The above statistics indicate that the world is fast becoming connected by internet and cyberspace activities, including Nigeria, and this by implication means a commensurate rate of cybercrimes and threats to digital economy. Cybercrime in Nigeria in particular has great implications for the digital economy because Nigeria is an emerging economy well-positioned to survive and prosper via e-commerce due to its population of about 200 million, enormous potential market, and diverse natural resources. The above assertion is supported by the statistics below.

The online purchase of consumer goods in Nigeria in 2020 discloses a significant number of users of digital economy in the country.

About 76.6 million Nigerians are shopping online, worth an amount of $3.37 billion; 2.2 percent of the GDP comprises the activities of online consumer. The value of the e-commerce market is recorded at $13 billion, which signifies an increase of 42 percent from 2019 and an average of $401 per consumer. Forty percent of transactions in 2020 were conducted using mobile phones. Of the payment method of e-commerce in Nigeria, the following may be said (as of 2020): 27 percent was credit card, 24 percent cash, 23 percent bank transfer, 9.0 percent e-wallet, and 17 percent other. The number of people who are making digital payment transactions is 105 million, with a total annual value of $21.61 billion, signifying an increase of 21 percent from 2019 with $206 as an average digital payment per digital payment user. The value of the total digital advertising market is $1.17 billion, with $412 million spent on digital search ads, $516 million on social media ads, $89 million spent on digital banner ads, $94 million on digital video ads, and $55 million on digital classified ads (Global Web Index 2020).

A summary of the views of the informants on the implications of cybercrimes on the digital economy in Nigeria is presented below in table 4.

Table 4: Summary of the Views of Informants on the Implications of Cybercrime on Digital Economy in Nigeria.

Category	Informants	Responses
A	1	Low level of digital trust
	2	Affect economic growth and development particularly e-commerce
	3	Prevents utilization of cyberspace for jobs creation
	4	Discourages entrepreneurs from initiating a business model on the internet
	5	Leads to loss of billions of USD
	6	Sabotage the economy
	7	Decline in the service sectors and Value Added Tax (VAT)
	8	Stymied modernization of business activities leading to physical transactions

B	1	Affects governmental revenue generation
	2	Consolidates overdependence on oil only as the major source of revenue
	3	Deprives service sectors from performing
	4	The usual relief of physical task of business is not benefitted from [*sic.*]
	5	Many businesses collapsed
	6	It deprives aspiring entrepreneurs of innovation and initiation
	7	Resources are wasted and lost unnecessarily
	8	Lack of confidence and cynicism about digital activities
	9	Secrecy is violated and compromised
	10	Waste of time, energy, and resources
C	1	Lack of digital trust
	2	e-commerce is crippled
	3	Cyberspace is being mis-utilized for social media only, without any benefit
	4	Energy and time is wasted in physical transactions
	5	Risk and uncertainties increased through physical transaction
	6	Valuable services are lost
D	1	People failed to benefit from digital transaction leading to low technological spread
	2	Mis-utilization of cyberspace for trivial activities instead of productive ventures
	3	Corporations are finding it difficult to operate nowadays
	4	Small scale businesses are not developing in e-commerce
	5	Investment is lost and wasted
	6	Public confidence is affected in e-businesses

Source: Field survey from interviews conducted between July to December 2018 with the selected informants by the researchers.

As before, some informants' major points are relevant to be quoted here. One of them disclosed that "the Nigerians are now enjoying digital economy through e-commerce and digital shopping from the comfort of their homes. But inadequate awareness, exposure, vulnerability, and weak cybersecurity is threatening the revolution in digital economy in the country even with the reality that the country can no longer survive without going digital, the colossal loss to hackers and cyber fraudsters recorded in the country disclosed that the digital economy has a serious challenge" (Interview with an informant in category C on August 6, 2018, at 10: 00 am). Another informant revealed that "corporations in Nigeria are finding it difficult to make transaction easier for its customers due to the alarming rate of cybercrime in the country and even the government itself is slow in transferring towards e-governance and e-transaction due to hackers and a weak cybersecurity in the country" (Interview with an informant in category A on August 17, 2018, at 3:00 pm).

Conclusion and Recommendations

The study concludes that Nigeria has recognized the importance of cybersecurity owing to its establishment of the Cybersecurity Act 2015, as a direct response to the recent emergence of cybercrimes in the country. The study contributed to the scholarly literature both empirically and theoretically. Empirically, the study revealed statistics regarding mobile and internet users in Nigeria as well as digital online activities in the country. It reported scientific views from informants on the effects of cybercrimes on national security and digital economy in Nigeria. Theoretically, the study succeeded in linking the connection between the various existing propositions on matters of national security and cybersecurity and the emerging empirical evidence from the area of study. The study also has policy implications in terms of initiating some proposals for policymaking in cybersecurity to prevent cybercrimes in Nigeria. The recommendations are listed below:

1. The government and other agencies responsible for ICT regulations should create popular awareness on the need to

prevent the exposure of security codes and other vital information online;

2. People should be sensitized to report suspicious transactions to the agencies that are responsible for the handling of cybersecurity policies;
3. Experts should be engaged to train existing staff within the relevant agencies on cybersecurity technicalities for easy identification of threats;
4. Workshops and media advertisements should be regularized quarterly to enlighten the public, organizations, and governmental agencies on the workings of cybercrime and how to adopt measures for protection;
5. Those who are caught in cybercrime activities should be punished severely according to the regulations of Cybersecurity Act 2015 to deter other criminals from future occurrence and
6. A databank for internet users and mobile phones should be created to enable the identification of suspicious calls and transactions in order for criminals to be apprehended immediately upon initiating an attack.

Acknowledgement of Funding

The work for this study was undertaken with research grant assistance from Professor Madya Bakri Bin Mat, an Associate Professor of Security and Strategic Studies in the School of International Studies, College of Law, Government and International Studies, Universiti Utara Malaysia.

References

Ajiji, Y.M. (2017). "Cybersecurity Issues in Nigeria and Challenges." *International Journal of Advanced Research in Computer Science and Software Engineering* 7(4): 315–321.

Alagappa, M. (1987). *The National Security of Developing States: Lessons from Thailand.* Dover, MA: Auburn House.

Alkali, R. A. (2010). *Issues in Nigerian Foreign Policy and International Relations.* Kaduna, Nigeria: Media Press.

Andreason, K. (2012). *Cybersecurity: Public Sector Threats and Responses.* London: Taylor & Francis.

Aniekan, M. N., & Afolabi, M.B. (2017). "Introduction to Cybersecurity and Cybercrime." In Aniekan & Afolabi (Eds.) *Intelligence and Security Studies Programme.* Lagos, Nigeria: Spectrum.

Antonucci, D. (2017). *The Cyber Risk Handbook: Creating and Measuring Effective Cybersecurity Capabilities.* Hoboken, NJ: John Wiley & Sons.

Azeez, O. (2019). "Cybercrime Cost Nigeria N288 bn in 2018." *Business a.m.* https://www.businessamlive.com/cyber-crime-cost-nigeria-n288bn-in-2018/, accessed March 25, 2020.

Bayuk, J. L., Healey, J., Rohmeyer, P., Sachs, M. H., Schmidt, J., & Weiss, J. (2012). *Cybersecurity Policy Guidebook.* Hoboken, NJ: John Wiley & Sons.

Beissel, S. (2016). *Cybersecurity Investments: Decision Support under Economic Aspects.* Geneva: Springer.

Buchanan, B. (2016). *The Cybersecurity Dilemma: Hacking, Trust and Fear between Nations.* New York: Oxford University Press.

Buzan, B. (1983). *People, State and Fear: An Agenda for International Security Studies in the Post-Cold War Era.* Chapel Hill: University of North Carolina Press.

Buzan, B., & Hansen, L. (2009). *The Evolution of International Security Studies*. New York: Cambridge University Press.

Cavelty, M. D., & Wenger, A. (2019). "Cyber Security Meets Security Politics: Complex Technology, Fragmented Politics, and Networked Science." *Contemporary Security Policy* 41(1): 1–28.

Chomsky, N., & Herman, E. S. (1979). *The Washington Connection and Third World Fascism*. Boston: Southend Press.

Chukwuma, O. A. I., & Mogom, I. A. (2018). "The Internet and National Security in Nigeria: A Threat-Import Discourse." *Covenant University Journal of Politics & International Affairs* 6(1): 20–29.

Collier, D. (1993). "The Comparative Method." In Ada, W. F. (Ed.), *Political Science: The State of the Discipline II*. Washington, DC: American Political Science Association.

Diogenes, I., & Ozkaya, E. (2018). *Cybersecurity Attacks and Defense Strategy*. Birmingham, UK: Packt Publishers.

Federal Government of Nigeria (2015). *Nigerian Cybersecurity Act 2015*. http://lawnigeria.com/LawsoftheFederation/Cyber-Crime-Act,-2015.html. Accessed Mar 25, 2020.

Fischer, E. A. (2009). *Creating a National Framework for Cybersecurity: An Analysis of Issues and Options*. New York: Nova Science.

Frank, I., & Odunayo, E. (2013). "Approach to Cybersecurity Issues in Nigeria: Challenges and Solutions." *(IJCRSEE) International Journal of Cognitive Research in Science, Engineering and Education* 1(1): 1–11.

George, A. L., & Bennett, A. (2005). *Case Studies and Theory Development in Social Sciences*. Cambridge, MA: Cambridge University Press.

Gerring, J. (2007). *Case Study Research: Principles and Practices*. Cambridge, MA: Cambridge University Press.

Global Cybersecurity Index (2018). *Global Cybersecurity Index*. https://www.itu.int/en/ITU-D/Cybersecurity/Pages/global-cybersecurity-index.aspx. Accessed March 25, 2020.

Global Cybersecurity Index (2019). *2019 Global Cybersecurity Index*. https://www.itu.int/en/ITU-D/Cybersecurity/Pages/global-cybersecurity-index.aspx. Accessed March 25, 2020.

Global Web Index (2020). *The Digital Trends to Watch in 2020.* https://www.globalwebindex.com/reports/trends-20. Accessed March 25, 2020.

Graham, M. (2019). "Changing Connectivity and Digital Economies at Global Margins." In Graham (Ed.) *Digital Economies at Global Margins*. Cambridge, MA: MIT Press.

Greenberg, L. T., Goodman, S. E., & Soo Hoo, K. J. (1998). *Information Warfare and International Law*. Washington, DC: National Defense University Press.

Guiora, A. M. (2017). *Cybersecurity, Geopolitics and Law*. London: Routledge.

Heintze, H., & Thielborger, P. (2016). "From Cold War to Cyber War: The Evolution of the International Law of Peace and Armed Conflict over the Last 25 Years—An Introduction." In Heintze & Thielborger (Eds.) *From Cold War to Cyber War the Evolution of the International Law of Peace and Armed Conflict over the Last 25 Years*, 3–9. Geneva: Springer.

Internet Crime Complaint Centre, FBI (2018). *Cybercrime Statistics: Top Ten Countries in the World.* https://www.ic3.gov/default.aspx. Accessed March 25, 2020.

Johari, J. C. (1997). *International Relations and Politics: Theoretical Perspectives*. New Delhi: Sterling Publishers.

Koops, B. (2016). "Megatrends and Grand Challenges of Cybercrime and Cyberterrorism Policy and Research." In Akhgar, B., & Brewster, B. (Eds.) *Combating Cybercrime and Cyberterrorism: Challenges, Trends and Priorities*. Geneva: Springer.

Kosseff, J. (2017). *Cybersecurity Law*. Hoboken, NJ: John Wiley & Sons.

Kremling, J., & Parker, A. M. S. (2018). *Cyberspace, Cybersecurity and Cybercrime*. London: Sage.

Kshetri, N. (2019). "Cybercrime and Cybersecurity in Africa." *Journal of Global Information Technology Management* 22(2): 77–81.

Leffler, M. P. (1990). "National Security." *Journal of American History* 77(1): 143–153.

Lijphart, A. (1975). "The Comparable-Cases Strategy in Comparative Research." *Comparative Political Studies* 8(2): 158–177.

Mazurczyk, W., Drobniak, C., & Moore, S. (2016). "Towards a Systematic View on Cybersecurity Ecology." In Akhgar, B., & Brewster, B. (Eds.) *Combating Cybercrime and Cyberterrorism: Challenges, Trends and Priorities*, 17–38. Geneva: Springer.

Maurer, T. (2018). *Cyber Mercenaries: The State, Hackers and Power*. New York: Cambridge University Press.

Maurushat, A. (2013). *Disclosure of Security Vulnerabilities: Legal and Ethical Issues*. Berlin: Springer.

Meeuwisse, R. (2015). *Cybersecurity for Beginners*. Canterbury, New Zealand: Icutrain Ltd.

Mohammed, K. H., Mohamed, Y. D., & Solanke, A. A. (2019). "Cybercrime and Digital Forensics: Bridging the Gap in Legislation, Investigation and Prosecution of Cybercrime in Nigeria." *International Journal of Cybersecurity Intelligence and Cybercrime* 2 (1): 5663.

Mordi, M. (2019). "Is Nigeria Really the Headquarters of Cybercrime in the World?" *Guardian*. https://guardian.ng/news/is-nigeria-really-the-headquarters-of-cybercrime-in-the-world/. Accessed March 25, 2020.

Mowbray, T. J. (2013). *Cybersecurity: Managing Systems, Conducting Testing, and Investigating Intrusions*. Indianapolis, IN: John Wiley & Sons.

National Bureau of Statistics (2018). Nigerian Population Estimate. https://www.nbs.org.ng. Accessed March 25, 2020.

Noam, E. M. (2019). *Managing Media and Digital Organisations*. New York: Palgrave Macmillan.

Ohwovoriole, O. (2019). "Nigeria Losses About N127bn to Cybercrime Annually." *This Day*. https://allafrcica.com/stories/2019061090144.html. Accessed March 25, 2020.

Olayemi, O. J. (2014). "A Socio-Technological Analysis of Cybercrime and Cyber Security in Nigeria." *International Journal of Sociology and Anthropology* 6(3): 116–125.

Omodunbi, B. A., Odiase, P. O., Olaniyan, N., & Esan, A. O. (2016). "Cybercrimes in Nigeria: Analysis, Detection and Prevention." *FUOYE Journal of Engineering and Technology* 1(1): 37–42.

Pelton, J. N., & Singh, I. B. (2015). *Digital Defense: A Cybersecurity Primer*. Cham: Springer.

Rayes, A., & Salam, S. (2019). *Internet of Things: From Hype to Reality*. Geneva: Springer.

Re-Hashed. (2019). *Cybercrime Statistics: A Closer Look at the "Web of Profit."*
www.thesslstore.com. Accessed March 25, 2020.

Reveron, D. S. & Mahoney-Norris, K. A. (2011). *Human Security in a Borderless World*. New York: Routledge.

Rittinghouse, J. W., & Hancock, W. M. (2003). *Cybersecurity Operations Handbook*. Amsterdam: Elsevier.

SaharaReporters. (2019). "Nigerian Government Directs NCC to Block Over Nine Million Unregistered SIM Cards."
http://saharareporters.com/2019/09/12/nigerian-government-directs-ncc-block-over-nine-million-unregistered-sim-cards. Accessed March 25, 2020.

Schjolberg, S., & Ghernaouti-Helie, S. (2009). *A Global Protocol on Cybersecurity and Cybercrime: An Initiative for Peace and Security in Cyberspace*. Oslo: E. Dit.

Schjolberg, S., & Ghernaouti-Helie, S. (2011). *A Global Treaty on Cybersecurity and Cybercrime*. Oslo: AiTO.

Schwanholz, J., & Graham, T. (2018). "Digital Transformation: New Opportunities and Challenges for Democracy." In Schwanholz, J. & Graham, T. (Es.) *Managing Democracy in the Digital Age: Internet Regulations, Social Media Use, and Online Civic Engagement*. Geneva: Springer.

Severo, M. (2019). "Safeguarding without a Record? The Digital Inventories of Intangible Cultural Heritage." In Romele, A., & Terrone, E. (Eds.) *Towards a Philosophy of Digital Media*, 165–182. New York: Palgrave Macmillan.

Shalhoub, Z. K., & Al Qasimi, S. L. (2010). *Cyber Law and Cyber Security in Developing and Emerging Economies*. Northampton, UK: Edward Elgar.

Sharan, M. P. (2009) *Qualitative Research Method*. San Francisco: John Wiley & Sons.

Singer, P.W., & Friedman, A. (2011). *Cybersecurity and Cyberwar: What Everyone Needs to Know*. New York: Oxford University Press.

Sule, B. (2018). *Political Party Financing and Election Reformations in Nigeria's 2015 General Election: Issues and Impacts*. PhD Thesis submitted to the School of International Studies, College of Law Government and International Studies, Ghazali Shafie Graduate School of Government, Universiti Utara Malaysia.

Sule, B. Yahaya, M.A. Rabi'u, A.A. Ahmad, M. & Hussaini, K. (2019). "Strategies of Combating Insurgency in Northeastern Nigeria: A Non-Traditional Approach." *Journal of Administrative Studies* 16 (2): 54–75.

Statista. (2019). "Nigeria: Number of Internet Users from 2017–2023 in Millions."
https://www.statista.com/statistics/183849/internet-users-nigeria/.
Accessed March 25, 2020.

Taylor, B. (2012). *The Evolution of National Security Studies*. Australian National University, National Security College Occasional Paper.

Touhill, G. J., & Touhill, C. J. (2014). *Cybersecurity for Executives: A Practical Guide*. Hoboken, NJ: John Wiley & Sons.

Walker III, W. O. (2009). *National Security and Core Values in American History*. Edinburgh, UK: Cambridge University Press.

Wolfers, A. (1964). *Discord and Collaboration: Essays on International Politics*. Baltimore: John Hopkins Press.

World Bank. (2017). *Combating Cybercrime: Tools and Capacity Building for Emerging Economies*. Washington, DC: World Bank.

Yakubu, Y. A. (2012). *Nigerian Foreign Policy from 1999–2012*. Zaria, Nigeria: Ahmadu Bello University Zaria Press.

Yin, R. K. (2018). *Case Study Research and Applications: Design and Methods*. London: Sage.

www.ingramcontent.com/pod-product-compliance
Lightning Source LLC
LaVergne TN
LVHW010719210926
848485LV00008B/1457/J
9781680537789